FEARFULLY
AND
WONDERFULLY
MADE

FEARFULLY
AND
WONDERFULLY MADE

THE ASTONISHING NEW
SCIENCE OF THE SENSES

MAUREEN SEABERG

ST. MARTIN'S PRESS
NEW YORK

First published in the United States by St. Martin's Press, an
imprint of St. Martin's Publishing Group

FEARFULLY AND WONDERFULLY MADE. Copyright © 2023 by
Maureen Seaberg. All rights reserved. Printed in the United
States of America. For information, address St. Martin's
Publishing Group, 120 Broadway, New York, NY 10271.

www.stmartins.com

Design by Meryl Sussman Levavi

The Library of Congress Cataloging-in-Publication Data
is available upon request.

ISBN 978-1-250-27241-6 (hardcover)
ISBN 978-1-250-27242-3 (ebook)

Our books may be purchased in bulk for promotional,
educational, or business use. Please contact your local
bookseller or the Macmillan Corporate and Premium Sales
Department at 1-800-221-7945, extension 5442, or by email
at MacmillanSpecialMarkets@macmillan.com.

First Edition: 2023

10   9   8   7   6   5   4   3   2   1

This book grew from a series of articles I cowrote with William C. Bushell, Ph.D., and our respective work on the senses. An MIT-affiliated biophysical anthropologist for twenty-six years, Bushell is one of the world's leading experts on adept meditators and humans with exceptional performance. He first pointed out to me the uptick in research into the senses.

Bushell has a savant-like, bird's-eye view of such things with more than forty years of delving into and building on the related literature. I'm honored that he has been my friend and mentor for more than a decade—together exploring the "cosmic, existential soup," as he puts it.

Thank you, Bill.

If the doors of perception were cleansed, everything would appear to man as it is, infinite. For man has closed himself up, till he sees all things thro' narrow chinks of his cavern.

—William Blake, *The Marriage of Heaven and Hell*

For I am fearfully and wonderfully made: marvelous are Thy works; and that my soul knoweth right well.

—Psalm 139:14

Contents

FEARFULLY
AND
WONDERFULLY
MADE

Introduction

The smell of the earth after rain.

The electric hues of a fiery sunset.

The zest of cardamom in a cup of chai.

Frissons along your cheek from the brush of
a hand.

The first few bars of your favorite song.

Smell, sight, taste, touch, and hearing. Together, these and other senses form our perception and give meaning to life. They do not just convey all that is beautiful; they are far more powerful than we know.

Contrary to lingering myths, humans perceive things as well as or better than other animals and any machine. Astonishing recent studies reveal that we can see light at the level of a single photon; hear sounds at amplitudes smaller than the diameter of an atom; use the vibrations of electrons to smell a trillion or more scents; taste fresh water;

and feel the difference of a single molecule of thickness on a smooth surface.

Our sensory systems, or sensoria, operate on an atomic and even subatomic scale. Yet we tend to think quantum mechanics is out *there* somewhere, in some faraway place where the tiny things of physics do their mysterious dance through space and time. Or we associate quantum phenomena with an underground tunnel in Switzerland where scientists work to find God particles. At the very least, *quantum theory* calls to mind a high-tech laboratory off-limits to everyday people, the wheelhouse of Albert Einstein and his colleagues. But the quantum is all around us and inside us; we are part of it.

What if we could understand the infinitesimal through our own experiences? And what if we recalibrated our expectations—our very lives—to match these newly discovered potentials? How far could we progress? Would the impossible become possible? Would our senses save us?

It is already happening, and people are making scientific breakthroughs one could call miraculous.

We need to develop what I call our *perception quotient,* or PQ, a type of intelligence fundamental to our survival. Central to the quotient is perceptual awareness—getting back in touch with and nurturing our abilities. The good news is that the senses are plastic and PQ can be expanded. After immersing myself in the research and many first-person testimonies for this book, I've put together a road map with ten clear steps to take. You can find that model, with the handy acronym PERCEPTION, in chapter 14.

I turned to people who can sense at superpower levels

to show the way. One stunning example is Joy Milne, a retired nurse in Scotland known for her ability to smell Parkinson's disease—she detected a change in her husband's odor six years before he was diagnosed with the devastating illness. Now researchers have found ways to use molecular biomarkers to detect the disease early on, which will allow doctors to intervene years before the worst symptoms set in. When I interviewed Milne, she revealed she is also able to smell COVID and is working with researchers on that incredible development. Milne's abilities were not recognized until later in her life when science caught up to these potentials. I'm sure there are more people like her waiting to be discovered and more people who could become like her through awareness and training.

The other senses are just as promising. In the case of vision, we are heading toward a day when human perception may unlock the ultimate secrets of the universe. Researchers recently proved that humans can detect a single photon of light with the unaided eye. The subjects in this study were not super-seers but three men with typical vision— one even wore contact lenses! Scientists are now working to discover if humans can see photons in superposition and entanglement. Will they see what has been theorized, or something else? We are on the brink of looking at the fabric of the cosmos!

Now that we are finding out about our true sensory natures, what else is possible? Maybe humans will discover a new kind of physics and make an enormous leap forward in space travel. Maybe doctors will learn how to smell cancer at the earliest stages and treat patients immediately. More children may develop superabilities if their education is based on the latest research. And, philosophically, by combining all these impressions and trusting them, maybe we will better understand who we are.

But there are a couple of opposing dynamics at work. Even as we learn we have far more sensory ability than we realized, humanity is becoming ever more desensitized in a troubled world. And North Americans and Europeans are spending too much time—as much as 90 percent of our lives—indoors, cutting ourselves off from nature, which dulls our senses. This disconnect explains why we have been so wanton with the entire ecosystem and ourselves.

As science advances and we find out how infinitesimal and fine the workings of our universe are, we must reconsider our place in it—our own bodies and our own sensory apparatuses. Have we not evolved within this wondrous and subtle and, yes, quantum system? Research is proving that humans are not separate from the universe, standing outside and looking in; we are "soft-tissue/high-tech" beings within it, according to anthropologist William C. Bushell.

Everyone, from the kindergartner to the trucker on a moonlit highway, has the potential to experience things on this level. In this book, I'll share the testimonies of study subjects and of folks outside of laboratory settings. I'll also share my own experiences as someone on the outer reaches of the senses. I am a synesthete and tetrachromat (super–color seer) and a subject of related research.

✳

How did this day arrive? In part, the massive investment in virtual reality, augmented reality, mixed reality, and robotics created a golden research era by funding perceptual laboratories. According to Statista, "Worldwide spending on robotics research is expected to increase from $55.8 billion in 2021 to $91.8 billion by 2026."

It is a two-way street. To create realistic experiences, researchers must measure human sensitivity with great precision. The irony is that the huge leaps in technology are upending the way we live and threatening to surpass us. But in this increasingly technological age, we can still celebrate the primacy of human biology.

In this time of tremendous change, we must reconsider our perceptual abilities. Humanity is at a crossroads. Some people advocate using implanted devices to become "transhuman." This is dangerous and premature, and people are still unaware of their own tremendous power and grace. I'd like to ring the alarm bell with one hand and hit the Pause button with the other until we are able to digest all the latest research pointing to our own extraordinary "equipment." Many things we seek may be possible through more natural, less invasive, means.

While I'm not anti-technology, I see this book as a celebration of what is still human and not yet machine. We have severely underestimated our abilities. Using the senses to their fullest potential is every human being's birthright.

Adept meditators have long reported sensing on these fine scales, and research shows that meditation improves all sensory abilities.

Our perception is tied to emotional health. Isn't it inter-esting how we say someone is "sensitive" when we're re-ferring to emotions? There is truth in this. Many people with heightened traditional senses are also exquisitely emo-tionally sensitive. It is time to create safe spaces for both emotion and perception. There is no such thing as "too sen-sitive." The prickliest among us are often justifiably so; they are the canaries in our collective coal mine, nature's early-warning devices.

The senses are the ultimate *doors of perception,* a phrase coined by poet and painter William Blake in his 1794 book, *The Marriage of Heaven and Hell.* They are gateways and, sometimes, portals of transcendence. Psychedelics are often used to achieve extraordinary sensory experiences, and the renewed interest in their value is a sign of a wider sensory renaissance. Yet the power to see the world in a new way is already in us.

I urge scientists to find more super-sensors among us so that their abilities can be marshaled to help the rest of hu-manity. The senses are still a largely untapped and precious health resource just beginning to be understood. They should begin their search among synesthetes, the 4 percent of peo-ple who have bonus senses, such as perceiving color when hearing music. Synesthesia is present in most super-sensors.

Running counter to the deluge of fear-based reporting on the impending "robocalypse," the evidence I present in the following chapters will, I hope, come as a relief and fill you with wonder and excitement. It's like finding out that the oil painting your grandmother willed to you is an old master. Ex-cept *you* are the painting. You are much finer than you know.

1

Olfactory Grammar

Humans are fond of pointing out the extraordinary sensory abilities found in the rest of the animal kingdom—an eagle's vision, a hound's ability to follow a scent trail. And we are not wrong, for nature is astounding. But we should no longer hold most of the sensory abilities of other creatures above our own. New research proves that our abilities are on par with the rest of the animal kingdom and beyond any machine.

Smell—or olfaction—is the detection and identification of airborne particles by the sensory organs. Humans can pick up perhaps a trillion odors. Yes, a *trillion*.

"The idea that human smell is impoverished compared to other mammals is a 19th-century myth," wrote Rutgers University neuroscientist John P. McGann in the journal *Science* in 2017. He traced this idea back to neuroanatomist Paul Broca, who in the 1800s hypothesized—incorrectly— that humans had small olfactory bulbs. McGann noted that the human olfactory bulb is actually quite large in absolute terms and that most mammals, including humans, have a similar number of olfactory neurons. In a much less enlightened age, the theory was also intended to separate humans from the rest of nature to show we have control over such animalistic impulses. The idea even influenced Sigmund Freud.[1]

"When an appropriate range of odors is tested, humans outperform laboratory rodents and dogs in detecting some odors while being less sensitive to other odors," McGann wrote in *Science*. "Like other mammals, humans can distinguish among an incredible number of odors and can even follow outdoor scent trails." He notes: "Humans' supposed microsmaty [literally, "tiny smell"] led to the scientific neglect of the human olfactory system for much of the 20th century, and even today many biologists, anthropologists, and psychologists persist in the erroneous belief that humans have a poor sense of smell."

No one in the world puts that myth to rest more than Joy Milne.

A retired Scottish nurse, Milne smelled the foul odor of illness many times during her long career. But a couple of years ago, she noted a particularly strange smell.

Milne was on a university campus, standing near a group of international students who had just returned from China. It was early 2020, a month before the first wave of COVID hit. And the smell was unmistakable.

"Apple cider plus infection," Milne told me.[2]

The tangy sweetness we associate with autumn harvest and hayrides was in the air, and it was an odd top note for the funk of sickness it accompanied. Milne's curiosity was piqued. The world had not yet descended into a pandemic. And Milne—a super-smeller whose abilities have been described by French perfumers as "somewhere between human and dog"—always picks up scents on people. She is a woman with a very developed perception quotient, or PQ, a type of intelligence I have coined that is fundamental to our survival. It is vital that we become more perceptive and get back in touch with and nurture our sensory abilities. She filed away the odd scent she picked up and reflected on it from time to time.

"Smell is different for me than most other people," Milne explained. "If somebody walks past me, I'm walking through the molecules they left behind for the next five, six minutes."

Then Milne caught COVID. When she smelled apple cider on her own skin, she realized what she had noticed in the students. The unthinkable happened: she temporarily lost her virtuoso gift. Like one-quarter of COVID sufferers, she could not smell. The virus attacks the chemosensory apparatuses in the human body, which are critically important for triggering an immune response. "It's like firing a guided missile into our bodies' 'radar facilities' in an act of war," William C. Bushell told me.

Fortunately for humanity, Milne recovered and inspired researchers to develop tests to detect coronavirus with skin swabs. That work is in progress at the University of Manchester.

Milne already had a global reputation due to her ability to smell Parkinson's disease. The dreaded illness, which attacks neurons in the brain, eventually leaves people unable to move and speak. Before Milne, there was no early diagnostic test. People could go ten years or more before the worst symptoms would appear and they'd already lost more than half of the related nerve cells. Milne noticed a distinct musky odor on her late husband, Les, six years before he had more severe symptoms.

At first, she assumed the smell was sweat from his long hours as an anesthesiologist. But her husband also began to suffer from profound fatigue. While attending a group meeting for people with the illness, Milne noticed they smelled musky, like her husband. She realized it might be significant and told scientists.

Researchers at the University of Edinburgh put her to the test. They asked Milne to smell twelve unmarked T-shirts, six of which had been worn by people with Parkinson's and six by people without the disease. They asked her to smell each shirt and determine whether or not it had been worn by someone with Parkinson's. She was correct in eleven out of twelve cases. And then, eight months later, the person she got "wrong" informed the team that he had been diagnosed with the disease. Now she had the scientists' full attention.

Milne's work with researchers at Edinburgh and the University of Manchester has now made early diagnosis of Parkinson's possible. With her help, scientists identified molecular biomarkers in sebum, a protective oily secretion on the skin.

How did they do it? They used a device known as a *mass spectrometer* to determine the chemical composition of the compounds that caused the odor.

"What we found are some compounds that are more present in people who have got Parkinson's disease and the reason we've discovered them is that Joy Milne could smell a difference," lead author of a related study, Professor Perdita Barran from the school of chemistry at Manchester, told BBC Scotland.[3]

Those compounds—hippuric acid, eicosane, and octadecanal—were found in higher-than-usual concentrations on the skin of Parkinson's patients.

"What we might hope is if we can diagnose people earlier, before the motor symptoms come in, that there will be treatments that can prevent the disease from spreading," Barran said.

Thanks to Milne's contribution, doctors may now have a ten-year lead on Parkinson's worst phases and can provide better early care. Her preliminary work on COVID is extremely promising, and she is also collaborating with scientists on cancer and tuberculosis detection.

Researchers are looking at sebum for clues about illnesses, Milne told me. Sebum is excreted by sebaceous glands and keeps the skin hydrated. Infants produce a lot of it. Sebum production declines in early childhood and picks up again in puberty.

Milne believes our bodies give us clues to our well-being all the time through these secretions; we just need to pay attention to them. "The best place to smell sebum is on the back of the neck running downward," she said. "There's always the best smell where the big sebum glands are. If it smells 'off,' the body is trying to send a message: *I'm ill. I've got something wrong with me. Could you please help?*"

Milne was not identified as a super-smeller until later in life. How many others like her are there?

David Howes is codirector of the Centre for Sensory Studies at Concordia University in Montreal. The center was founded in 1988, well before most people were interested in the senses, when sociologist Anthony Synnott and Howes won a grant to research "the varieties of sensory experience." They were joined by Constance Classen, then a doctoral student at McGill University and now an internationally known researcher on the culture of the senses. Howes is the author of many books on the senses, including *Aroma: The Cultural History of Smell*, with Synnott and Classen.

Howes is concerned that the world is becoming more

complex technologically and more homogenized sensorially. "The very drive toward making fewer sensations is going to precipitate a counterreaction," he told me.[4] "For example, perfumes and so forth are all very sweet now. There are a few exceptions, like Opium, but they're not really moving the parameters."

In fact, he thinks humans are odor-phobic. "Witness all of our deodorizing and then re-odorizing rituals: the morning shower followed by adding all these artificial scents," he told *The Guardian*. "Our noses are woefully uneducated now, and I'm very much in favor of liberating the nose. It has been kept down for too long."

Super-smeller Milne believes the key to developing the sense is paying attention. She told me she intentionally inhales a smell at least twice, at different stages of its unfolding. Now an honorary lecturer at Manchester, she recently collaborated with the BBC on a program to teach children to use their sense of smell better.

"If you encounter something and you leave," she explained to me in a Zoom meeting, "and then just not sniff it again, you have sniffed the one smell. But if you think about what you're letting go into your brain and then you sniff again, the molecules will have started to disperse, and you get a rounded smell. You are beginning to collect a smell."

She teaches kids to be more present and mindful of their senses. "You've got your nose with your sensors and sensory endings. But you have the nerve system which goes back up into the brain, and it splits. So on the one hand, you get the sense of smell immediately. On the other hand, it is saved in an olfactory library."

For Milne, scents are accompanied by synesthesia *photisms*—colored shapes only she can see. This makes the experience more memorable—it's like adding fireworks. Researcher Eric Haseltine, author of *Long Fuse, Big Bang*, determined she had synesthesia after interviewing her at length in 2018.

"Surely other super-smellers had been exposed to Parkinson's disease over the last several decades, but Joy was the first one to notice a distinct odor associated with the disease and to report her experiences to biomedical researchers," Dr. Haseltine, a sensory neuroscientist, reported in *Psychology Today*.[5]

Yes, she has an exquisitely sensitive nose, he wrote, "but she apparently also employs non-olfactory parts of her brain—the visual regions of her cerebral cortex—to process olfactory sensations. Possibly, one reason Joy excels at detecting Parkinson's disease is that she engages more neurons (i.e., visual neurons) when smelling odors than 'normal' people."

Dr. Laura Speed and her colleagues at Radboud University in the Netherlands conducted a study of synesthetes that supports Dr. Haseltine's hypothesis.[6] "Dr. Speed's team found that test subjects who experienced colors whenever they smelled odors outperformed 'normal' subjects at discriminating subtle difference in both odor and color," he said.

More proof of this theory is evident in the extraordinary sensory experiences of synesthete and painter Carrie Barcomb from Brandywine Valley, Pennsylvania. A few years ago, a friend told Barcomb about Joy Milne's ability to smell illness, and Barcomb, a mother of four, had an epiphany.

She realized she had been smelling sickness all along: the sour smell on her children's breaths when they had stomach viruses; the burning scent in her mouth when she started to get a cavity; the musty odor of her pet dog at the beginning of a flea allergy.[7]

Barcomb then worked to understand and master her ability. Looking back on her life, she said, it was there all along. As philosopher Søren Kierkegaard observed, "Life can only be understood backwards, but it must be lived forwards."

Barcomb approaches her talent with verve, and she's already teaching the next generation. You might find her in the family kitchen asking her kids to identify spices without seeing the labels. Like Milne, she believes these skills should be cultivated early.

Barcomb told me when I interviewed her for *Psychology Today* that the sense of smell is like a language.[8] "You learn the smell of rain, snow, or electricity in the air from lightning," she told me. "The fungus in the earth in the springtime is much more pleasant compared to the leaf mold of autumn."

She is so sensitive to odors that she says she must "make friends" with them and allow her body to adjust. "Sometimes I must try to turn it off, dull it. With anything toxic (gasoline, cigarettes, turpentine), I must hold my breath to not smell." Barcomb had to warm up to cheese too. "As a child I was unable to be around cheese! (Everyone thought I was weird.) I finally accepted cheese at age 25. With my nine-year-old daughter, it's the same. I have made it a point never to insinuate anything negative about it."

Carrie Barcomb represents to me what I hope will

happen globally when the testimonies I present here gain traction in our human conversation. Think of me as the friend who tells you, as her friend told her, that these things are possible.

One of the pioneers of olfactory research, Charles Wysocki of the Monell Chemical Senses Center in Philadelphia noted that olfactory sensitivity runs in families. And due to a happy accident in his laboratory, he discovered people can improve their sense of smell.[9]

Wysocki was trying to determine why some people are unable to detect specific scents. He focused on the compound androstenone. Fifty percent of people cannot detect this smell, which is similar to urine. According to the Monell Center, Wysocki and his colleagues demonstrated that identical twins, who have virtually the same DNA, were identical in their ability to smell (or not smell) this chemical, indicating the trait is most likely genetic.

Wysocki was one of the people who could not detect androstenone, so he volunteered to prepare this unpleasant solution for the research. However, after a few months of working with the compound, something astonishing happened. He began to smell the androstenone! This serendipitous observation paved the way for recent research about how one's environment fine-tunes the sense of smell.

Different creatures experience the world in different ways. Humans can't see infrared and ultraviolet light like a mantis shrimp can; we can't hear a rodent burrowing underground the way an owl can or track the scent of a murderer with the skill of a bloodhound. But as Milne and Barcomb demonstrate, our species' perception of things—

the top-down coalescence into reality—is superior in many ways to other animals'. And they, along with Wysocki, prove it is a plastic ability that can be learned and grown.

Of all the senses, the sense of smell has been with animals the longest. "Chemo detection—detecting chemicals related to smell or taste—is the most ancient sense," Dr. Dolores Malaspina of Columbia University told *Everyday Health*.[10] "Even a single cell animal has ways to detect the chemical composition of the environment."

The power of smell has led to marketing efforts to scent everything from the sidewalks outside Dunkin' Donuts with a coffee aroma to the American Express skybox at Lincoln Center with a signature scent in an explosive wave of "multisensory marketing." Essentially, things perceived by two or more senses are more memorable.

The potential for human olfactory detection may be almost infinite. In March 2014, the journal *Nature* reported that scientists at The Rockefeller University, in New York City, determined that people can smell at least a trillion odors.[11]

This was challenging to prove, because, unlike wavelength for vision or frequency for sound, the parameters of smelling are not fixed. Also, scents can vary ever so slightly, and cataloging them is difficult.

"It's hard to organize odors," Donald Wilson, an olfactory researcher at New York University School of Medicine, told *Nature*. "How does one compare the musky scent of a

U.S. drugstore cologne like Axe body spray with a rival, such as Old Spice, or with something that smells of vanilla?"

The scientists made mixtures of various odorous molecules. They gave subjects three vials, two of which had the same mixture, and asked them to identify the one that was different. They found that people could determine which vial held the different scent as long as the components of that mixture differed by more than 50 percent from the mixture in the other two vials. The scientists then calculated the number of possible molecular combinations that had an overlap of less than 51 percent—and could thus be smelled by humans—and came up with one trillion. A coauthor of the study, Andreas Keller, said, "My hope is that this helps to dispel the myth that humans have a bad sense of smell."

Like Milne and Barcomb, Debra Pollicino-Ludwig, a pediatric nurse in New York City, busts that myth wide open. She can identify concerning scents on the skin of newborn babies in her care.

"The first thing I do when I take an infant into my arms at the hospital is close my eyes and smell," she told me.[12] She can spot early signs of congenital illness in her charges or issues with the mom. "I can smell if the mother had prolonged rupture of membranes and is beginning to develop chorioamnionitis [an infection of the membranes and amniotic fluid around a fetus] even if it hasn't been 'called' by the obstetrician. I sometimes get overwhelmed by the smell of urine on the baby if the mother urinated during delivery. I could smell on the baby if the mother received antibiotics during labor. I describe a healthy newborn as smelling like modeling clay, which I find a little funny." Pollicino-Ludwig said she is so sensitive that she can tell if the mother has recently eaten garlic, onions, or fenugreek. In one case, she recalled, "I could smell maple syrup on a newborn during my initial assessment after delivery. I remembered learning of a very serious condition called *maple syrup urine disease.* It is a metabolic disorder. The baby is missing enzymes to break down milk. The baby cannot even have breast milk. I asked my friends to smell it, and they couldn't smell it, but every time I moved the baby, I could smell maple syrup. I told the attending, who called the neonatal intensive care unit to assess the baby."

After a series of tests on the infant, the newborn's mother recalled that she'd eaten a lot of fenugreek leading up to delivery because she believed it would promote milk production. Fenugreek smells like maple syrup. The mother was very grateful to the alert nurse and relieved her baby would be okay.

Pollicino-Ludwig has always been sensitive to the scent of disease, everything from a fungating breast tumor to strep throat. "I can smell the notes in perfumes and tell you what a lot of the ingredients are. I have olfactory hallucinations as an aura before a migraine. I usually smell antifreeze or ammonia."

It is not surprising that Pollicino-Ludwig is also a synesthete. The hyperconnectivity of synesthetes' brains seems to lay the groundwork for enhanced senses, as you'll see many times in this book. And synesthesia appears to be a latent ability in all people.

Lesley Roy of Florida is a vibrant, active woman who is enormously sensitive to smell, yet she is not a synesthete. About fifteen years ago, she noticed a change in the smell of her urine and her breath.

"It was incredibly distinct, like a newly discovered chemical. I thought it was strange, like after eating asparagus. I never had my tonsils out, so these flesh sacks that supposedly serve no purpose may have been tipping me off that something wasn't right. They had always been speckled in times of stress with tiny white dots—a sign my body was trying to expel toxins."[13]

Roy didn't want to admit the nagging feeling she had: *This is what cancer smells like.*

It took many tests and doctors to confirm her worst fear, and it took fourteen years to recover. Roy said she has never told anyone, not even her doctors, that it was the change in the smell of her urine and breath that made her seek testing; she was otherwise mostly asymptomatic at the time.

"I know now that it is critically important to listen to your body with all your senses, to even the smallest of changes, to tune in and hear (or, in this case, smell) what it is telling you."

How exactly does smell work? A long-held theory, known as the *lock and key model,* proposes that specific odor molecules fit into and activate specific odor receptors.

Dr. Luca Turin, of the Alexander Fleming Biomedical Sciences Research Center in Athens, Greece, is among many biophysicists who believe there is a different explanation. His theory is that a molecule's vibrations, not its shape, determine its odor.

Turin does not believe we have the ability to smell a trillion or an infinite number of scents. He thinks it is impossible to tell, though he confirms that it is "a lot."

When I interviewed him, he explained the debate.[14] Essentially, it comes down to the fact that odor receptors are very nonspecific, so the idea that odor molecules must activate certain receptors is unlikely. The vibration theory holds that a molecule's vibrations, not the receptor that the molecule binds to, is what determines its odor. Turin found a way to test this theory. First, he noted that thiols and boranes (two different compounds) smell the same to humans—like sulfur—despite not sharing any chemical properties. What they do share is a vibrational frequency. Turin then turned to fruit flies, which have an entirely different olfactory system from humans and so should not interpret these two very different compounds as having the same smell.

What he found was that when fruit flies were trained to avoid boranes, they also avoided thiols, and vice versa. This suggests that the fruit flies were responding to the vibration of the molecules, not their shape. Smell is happening on the quantum, vibrational level.

Smell is not just the nose's job; there are olfactory receptors in many body organs. Researchers Désirée Maßberg and Hanns Hatt reported in *Physiological Reviews* in 2018 that they had found olfactory receptors "in all other human tissues tested to date."[15] These include the "testis, lung, intestine, skin, heart, and blood." While the receptors and their functions are still being studied, a leading theory is that they act as an early-warning system against disease. How many of us have a sense of smell like the people in this chapter but don't talk about it because it sounds too far-fetched? How many of us just dismiss these impressions as we go about our busy lives?

These impressions are legitimate. Knowing this is the first step on the journey to realizing your potential.

My hobby is nature photography. I think of the flora and fauna I encounter outdoors as parallel nations of sentience. While working on this chapter, I learned that a litter of baby foxes had been born in one of the forests where I walk. I knew the general area, but there were no major landmarks to guide me in a section thick with trees. Inspired by the incredibly high PQ of the women I was writing about, I decided to use my senses. I closed my eyes and took a deep sniff. There it was—the unmistakable odor of wet dog. I opened my eyes and followed it for a while until I spotted freshly disturbed soil in the distance. Even I could follow my nose to the den. I stood a respectful distance away and watched three beautiful kits at play as early-morning sunlight streamed down to the woodland floor. It was incredibly moving, and I've never felt wilder myself. It is moments like this that I wish for you, readers.

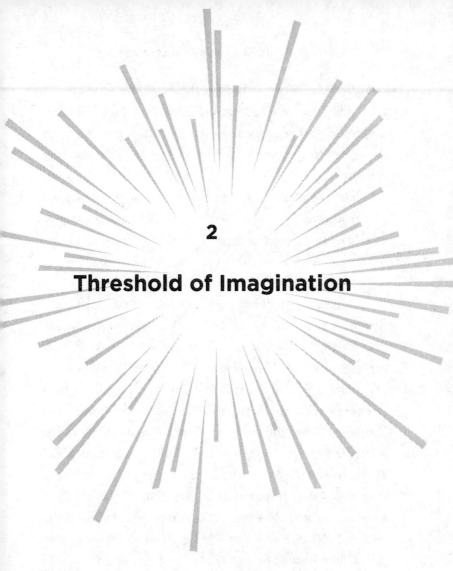

2

Threshold of Imagination

Four hundred years ago, Johannes Kepler, Galileo Galilei, René Descartes, and Isaac Newton made the astronomy and physics discoveries for which they are widely known, but these giants also had papers on their desks filled with notes and drawings of the human eye. The seventeenth century was a time of passionate inquiry into optics.

Kepler published the first explanation of the mechanism

of vision. Galileo described his theory of the senses in his 1623 *Il Saggiatore* (*The Assayer*).[1] Descartes discovered that the lens of the eye focuses light on the retina. And Newton's prism experiments were fundamental to our understanding of color.

These geniuses of the early modern era are better known for the enormous strides they made in unlocking the universe's secrets than for their enthusiastic study of vision. But it is appropriate that this era, which gave us telescopes and microscopes, also led to great leaps in eye research.

"Reflections on the senses, and particularly on vision, permeate the writings of Galileo Galilei," wrote Marco Piccolino in the journal *Perception*.[2] But, he noted, "this aspect of his work has received scant attention by historians."

In Galileo's time, people thought that the objects they interacted with generated the senses they experienced— that is, the sweet fragrance was in the rose itself, not in the human olfactory system; the feel of silk rested in the luxurious material's fibers, not in the sense of touch. That's almost as absurd to us now as the idea that the sun revolved around the Earth. But just as Galileo proved the sun is the center of our solar system, he believed what we saw, heard, tasted, and touched came from our bodies, not external objects. He reoriented our thinking of both the universe and ourselves in crucial ways.

"As far as concerns the objects in which these tastes, smells, colors, etc., appear to reside, they are nothing other than mere names, and they have their location only in the sentient body," he wrote in *Il Saggiatore*. "If the living being were removed, all these qualities would disappear."

Galileo wrote that without a living ear, there would be no sound in the world, "only vibrations"; without a living eye, there would be no light, no color, no darkness, "only the imponderable oscillations that correspond to light and its matter, or their absence."

This shifting of perception from the object to the individual was as significant as his proving that the sun, not the Earth, was at the center of the solar system, and it led, eventually, to the modern view of sensory processing. Even as they were looking skyward, these historic figures, particularly Galileo, knew they must also look inward to understand the nature of things.

Other people all over the globe intuited this as well. From ancient times, people of many cultures have sought to explore reality through contemplation. Some adept meditation practitioners experienced advanced capabilities of all the senses, including vision, while pondering the infinite. Tibetans, Chinese, Israel's Essenes, the Egyptians, the people of Colombia in South America, the Greeks, and others had such experiences. Often when they retreated to dark environments in their meditations, they saw tiny bits of light, according to William C. Bushell.[3]

"These latent human capacities have been recognized and cultivated by elite practitioners in several non-Western cultures as a means of systematically investigating the physical, including quantum physical, nature of the universe," he told me.

Physics and eyesight mingled for centuries, but in 2016, researchers at the Rockefeller University in New York married them for eternity in one tiny, brilliant, and historic flash. They proved humans could see light at the level of a single photon—the smallest particle of light—with the unaided eye in laboratory conditions. The connection between vision and physics intuited long ago was now undeniable.

Photons are massless bits of radiant energy. To understand light, first think of electrons orbiting an atom's nucleus. Electrons have natural orbits they occupy, but you can move them to higher orbits by energizing the atom. When an electron falls from a higher-than-normal orbit to its usual one, it emits a packet of energy—a photon.

Western science has known humans can detect light at very low levels (five to seven photons) for seventy years. But scientists only recently developed a single-photon light source that could demonstrate the power of our eyes.

Experimenters had the light source deliver single photons to the eye. They found that subjects sensed the slightest glimmers. The particles created by falling electrons met the vision receptors in the retina.

Somewhere, Galileo smiles.

Physicist Alipasha Vaziri conceived of and led the experiments at Rockefeller. He read early papers on light detection from the 1940s and was fascinated. "Human senses have evolved to be in touch with the smallest portions or quanta of the physical world: touching single molecules, seeing photons," he told me.[4] Humans are macroscopic organisms made up of billions of cells, and yet they are sensitive enough to interact with something as small as a quantum of light.

To illustrate this, let's consider the size of a photon compared to the size of the rod, the retina cell that perceives it.

The photon is one hundred thousand times smaller than the cell! And as Vaziri pointed out, each of us is a confederation of billions of cells.

For humans to see, light must pass through the cornea (the protective layer at the front of the eye), then through the pupil, which controls the amount of light entering the eye. From there, light goes through the lens, a biconvex (rounded on both sides) body that focuses light on the retina. The retina, a light-sensitive membrane at the back of the eye, contains photoreceptors that turn light into electrical signals, and those signals travel along the optic nerve to the brain.

Rods are a type of photoreceptor sensitive to light. Each retina has approximately 120 million of them; they help you see at night and are essential for peripheral vision. You're familiar with how your rods work—they take over from the cones when you, say, go into a darkened movie theater, and they need seven to ten minutes to warm up. You experience this as difficulty seeing at first, then a slow improvement in vision as your eyes adjust to the dimmer light.

But *detecting* a single photon is not the same as seeing it. The most amazing thing, Vaziri said in a very poetic statement to the *Los Angeles Times*, is that "it's not like a dim flash of light or anything like that. . . . It's more a feeling of seeing something rather than really seeing it. Like being [at the threshold] of imagination." He compared it to looking at a faint star in the night sky: "One second you see it, but the next second you don't."[5]

He elaborated for me: "I was going in with the assumption that I would be seeing some kind of a faint light, and I would try to do my best to see, and the light we are using is green, so I would be seeing green."

But in reality, he said, "it didn't have any color per se . . . You are not even sure you saw something."

In the experiment, each subject was placed in a completely dark room. To ensure that the light stimuli would reach the area of the retina with the most rod cells, the subject's head was kept in position by a headrest and a bite plate. The participant sat in the room for forty minutes to let his eyes adjust before beginning the test.

He then pushed a button and heard two sounds separated by one second. Sometimes a photon was emitted at the same time as one of the sounds. The subject then had to respond with one of two options: yes, he had seen a light, or no, he had not. He also had to rate how confident he was in his answer.

Vaziri and the three male graduate-student subjects who were tested over many thousands of trials answered correctly a statistically significant number of times. Two of the young men had average vision, and the third wore contact lenses. Vaziri wears glasses. None of the men were super-seers. Yet there was a "knowing" that is undeniable in the results.

Further, when the subjects were surer of their answers, they were more likely to be correct. They were also more likely to see a photon if it had been preceded by an earlier photon, which suggests that the first photon primed the visual system.

As momentous as this research is, it raises other avenues of inquiry. Vaziri was trained as a quantum physicist, and for him, this brought up another fascinating question: Why has evolution pushed organisms to be able to interact with a single quantum of light?

Rebecca Holmes, a physicist at Los Alamos National Laboratory in New Mexico, is currently working on ways to use photons to create "license plates" for the many satellites in our crowded skies. She, too, worked on detecting light particles during her doctoral training at the University of Illinois.

Writing for *Aeon*, Holmes reflected on those years. "I spent a lot of time in the dark in graduate school," she said. She was studying quantum optics and how humans perceived the smallest amounts of light, and, she said, "I was the first test subject every time."[6]

The experiments were conducted in a closet-size room on the eighth floor of the psychology department at the University of Illinois. Special blackout curtains and a sealed door ensured that the room was in total darkness. "For six years," she wrote, "I spent countless hours in that room, sitting in an uncomfortable chair with my head supported in a chin rest, focusing on dim, red crosshairs, and waiting for tiny flashes delivered by the most precise light source ever built for human vision research. My goal was to quantify how I (and other volunteer observers) perceived flashes of light from a few hundred photons down to just one photon."

Holmes, who is also a talented poet and musician, talked with me from Los Alamos about the experiments and shared an excerpt from her Ph.D. dissertation about what it is like to encounter photons.[7]

"The dimmest flashes we studied (apart from the true

single-photon trials, which are still ongoing) deliver about thirty photons to the eye, of which about three are detected by rod cells," she wrote. At that level, subjects couldn't really be sure if they'd seen anything; they just reported a feeling about which side of the visual field seemed "different."

"In the clearest trials, participants might perceive a slight motion to one side or a tiny suggestion of a flash. The flash itself is sometimes small and localized, like a star in the sky, and sometimes larger and diffuse. It has no color. In total darkness it becomes difficult to judge exactly where in the periphery the flash originated, beyond left or right, higher, or lower."

Compared to single photons, she wrote, "Flashes of three hundred photons, of which about thirty are detected, are strikingly easier to see."

Vaziri, you'll recall, pointed out the enormous difference in scale of the tiniest portion of light to the billions of cells in the human body. Holmes explained their nature in terms of energy. "Even these brighter flashes deliver less than 75 electron volts to the retina," she wrote. "A flying mosquito has about sixteen billion times more kinetic energy!"

She had another critical observation that confirmed Vaziri's findings. With practice, people seemed to get better at detecting photons.

"One thing that we found in our photon-vision research was that some people are better at it than others, and some people get better with practice," she told me. "Some people just seemed to be better, maybe more sensitive, or better at paying attention." She wondered if it was improvement based on repetition or if they were "actually tuning their visual system to perform really well at this detection task."

This reminded me of a story retired U.S. Navy captain and Fordham University law professor Lawrence Brennan told me. In World War II, the Japanese military trained pilots to see stars in broad daylight to better navigate with their eyes, and they, too, improved with practice.

In his autobiography, *Samurai! The Personal Story of Japan's Greatest Living Fighter Pilot,* Saburo Sakai wrote, "One of [the pilots'] favorite tricks was to try to discover the brighter stars during daylight hours. . . . Our instructors constantly impressed us with the fact that a fighter plane seen from several thousand yards often is no easier to identify than a star in daylight. And the pilot who first discovers his enemy and maneuvers into the most advantageous attack position can gain an invincible superiority. Gradually, and with much more practice, we became quite adept at our star-hunting."[8]

Then they went even further. "When we had sighted and fixed the position of a particular star, we jerked our eyes away ninety degrees and snapped back again to see if we could locate the star immediately. Of such things are fighter pilots made."

Others have looked into the sky and seen what almost no one else can see. "When the skies are clear, and the moon is not too bright," wrote Bill Bryson, "the Reverend Robert Evans, a quiet and cheerful man, lugs a bulky telescope onto the back deck of his home in the Blue Mountains, about fifty miles west of Sydney, and does an extraordinary thing. He looks deep into the past and finds dying stars."[9]

In his bestselling *A Short History of Nearly Everything*, Bryson wrote of his visit to Robert Evans, a man who hunts—and finds—supernovae with nothing but his eyes and a telescope. The late neurologist, author, and physician Oliver Sacks called him a savant in his book *An Anthropologist on Mars*.

How difficult is it to pick out a new supernova in the night sky? Bill Bryson explained it this way: "Imagine a standard dining room table covered in a black tablecloth and someone throwing a handful of salt across it. The scattered grains can be thought of as a galaxy. Now imagine fifteen hundred more tables like the first one—enough to fill a Wal-Mart parking lot, say, or to make a single line two miles long—each with a random array of salt across it. Now add one grain of salt to any table and let Evans walk among them. At a glance, he will spot it."

Savant Jason Padgett suffered a traumatic brain injury in a mugging, and since then, he has been able to see what he believes is the fabric of things. We coauthored the book *Struck by Genius* about his extraordinary journey from college dropout to someone who can comprehend complex math and physics and has even proposed his own theories.

Padgett sees discrete geometric forms all around him in a visual synesthesia response to math and physics concepts. For example, pi is a circle made of triangles. He draws these shapes to help him do complex calculations. Padgett had never taken anything more advanced than an algebra class before his brain injury. He seems to be living proof of our ability to discern the infinitesimal.

"The universe is quantum (made of discrete indivisible pieces) at the smallest and most fundamental level," he told me recently.[10] "As our species has evolved, our DNA has adapted over time. Since our evolution is directly affected by our environment and that environment is quantum at its core, it stands to reason that our senses are also quantum at the most basic level."

When we see light, he said, "it is coming at you in tiny discrete quantum packages. It is all information, and our brains and DNA must evolve according to quantum laws."

The late Dr. Darold Treffert, the world's leading savant authority, worked with both Jason Padgett and the amazing savant Kim Peek, the man on whom the film *Rain Man* is based. I interviewed Treffert in his office in Wisconsin. "You must remember, Maureen," he said, "that nothing was added nor created when Jason Padgett was injured. Rather, factory-installed software, genetic memory, was released." Treffert explained that while these processes had risen to the conscious level in Padgett, they are within all of us.[11]

It makes me wonder how we can all be exceptional and yet still live in ignorance of our power.

In fact, Charles Darwin was awestruck by the perfection of our eyes. They are so extraordinary that he admitted it seemed "absurd" they had evolved through natural selection. Science is proving that our eyes are even more impressive than Darwin imagined.

✳

How did our eyes become so magnificent? Their story began four billion years ago with the very first single-celled organisms. These oceanic creatures were not exactly able to see, but they could sense light with simple eyespots.

The organisms dove deep into the sea to avoid the harsh ultraviolet light of midday and surfaced at the end of the day to photosynthesize in gentler orange light. Today's single-celled organisms still have eyespots with photoreceptors that detect the direction and intensity of light.

The molecule the earliest life on Earth used to sense light is rhodopsin. And in an unbroken thread through billions of years, the same molecule evolved into the light-sensing pigment we humans use now for color vision.

About 540 million years ago, a huge number of organisms began to appear on the planet in an event known as the *Cambrian explosion*. It is not a coincidence that eyes similar to our own first appeared during this era. Vision was a driving force in the vast diversification of the fauna, allowing them to see and capture prey. Vision was essentially a superpower, and once the eye evolved and animals could hunt, the planet changed forever.

Sensory researcher Kristopher Jake Patten of Arizona State University studies my eyes because I have a mutation that gives me super-color vision. He explained, "The eye picks up on the visible spectrum of light not because it's an arbitrary range of electromagnetic radiation that an eye randomly decided to see, but because it corresponds to the peak of the black-body radiation given off by our sun." The organisms that first walked on land that could see in this range of light had better vision than those that didn't, so this mutation was the one that got passed on, and now all terrestrial animals see in that range. "We and other organisms are all part of an intricate dance with our environment," he said.[12]

And what a complex dance it is. "As a predator's legs evolve to hunt faster, so, too, must their eyes and visual cortex evolve to make sense of the speed of their movement," Patten said. "As plants evolved poisons to keep animals from eating them, the olfactory and gustatory systems had to evolve to detect these chemicals. In both cases, the physical environment contains valuable information that's available if only a species can evolve (through random variation in sensory capabilities) to pick it up."

Only recently has science had the power to measure the upper limits of our visual capabilities. How powerful are eyes? Their function takes up a full 60 percent of the brain's cortex. The muscles of the eyes are more active than any other muscle in the human body. The retina is as neuroplastic as gray matter.

Where could this all lead? Possibly to humans directly observing the universe of quantum mechanics.

Two of the fundamental principles of quantum mechanics are entanglement and superposition. *Quantum entanglement* means that two or more particles are linked; they act as if they are joined together no matter how far apart they are in space. If something happens to one of the particles, the other particles in the system are affected wherever they are. *Quantum superposition* means, essentially, that an electron can be in two places at once. Photons are quantum particles. If humans can see photons, maybe there are ways to directly observe the quantum universe.

Particles that are in two places at once are not an

observable phenomenon, but according to Project Q at the University of Sydney, which investigates how related technologies will change the world, "If it were, our understanding of what we understand to be 'real' would be challenged, opening the door for a whole host of quantum weirdness that classical theory keeps at bay."[13]

Biophysical anthropologist William C. Bushell believes that adept meditators would be really good subjects for the next round of research into the eyes. Studies conducted by Vaziri, Holmes, and others indicate that research subjects' ability to control attention, patience, and bodily stillness is crucial for the studies' results, and experienced meditators have all these qualities.

Bushell points out that Tibetans described lightproof chambers thousands of years ago in which they saw light at this level. "Monasteries, through history, have really been laboratories, and the monks' goal has always been 'yogic direct perception.'"

Some advanced practitioners who engaged in forms of "eyes-open" meditation have already suggested what may be descriptions of the extraordinary quantum phenomenon known as *superposition*. As Bushell relayed to me, "In such accounts by these meditators, objects may appear blurry, 'cloud-like,' and/or existing in more than one location simultaneously. . . . And along with superposition, there is evidence that the human eye might be capable of discerning quantum entanglement in light, according to recent studies demonstrating the capacity to discern different states of light *polarization*—or light in a single plane."

We've come a long way since men investigating the

cosmos covered their desks with drawings of eyes. It turns out that vision and the cosmos—indeed, all our senses and the cosmos—have always been interrelated. Some of the greatest thinkers in history intuited it, and now some of our finest contemporary minds are proving it in laboratories.

Physicist Carl Sagan said, "We are made of star-stuff. We are a way for the universe to know itself."

I believe that in the grand cosmology, we are the universe's sentience.

3

Peregrinations Through Sensations

I live in a sensory frontier as a pilgrim of perception. Scientists are only beginning to explore the neurological territory of synesthetes and tetrachromats (super–color seers) like me. I spent almost half my life alone on this island of sensation before explorers arrived. I'm glad they're here. And I'm thrilled to be alive during this renaissance of research into the senses.

Sometimes I hear sounds for things that move silently to others. For example, a popular GIF of two pylon towers jumping rope makes me hear *thud, boing,* even though there's no accompanying soundtrack. Sometimes I get tingles on my cheeks or forearms in response to high-pitched music, such as a soprano singing. I am a polysynesthete, meaning that I have many bonus senses, including seeing colors associated with numbers, letters, music, days of the week (Tuesday is golden), months, and other stimuli. These impressions do not mean I don't also experience the primary expected sense. I just get a layer on top of it. One example is mirror-touch synesthesia (MTS), and it has nothing to do with color. I have a preponderance of mirror neurons in my brain, which make me feel the pain or pleasure of other people. Everyone has these neurons, but 1.6 percent of people have the related synesthesia, resulting in profound and literal empathy. I also have tetrachromacy—four types of cones for seeing color rather than the usual three—so I can see exponentially more color than most people. I am a super–color seer, and it's possible that synesthesia is related to my supersensory abilities. The members of my immediate family do not share these traits, though a couple of my cousins are also synesthetes.

To say I am a neurological outlier is an understatement. But there is a certain objectivity to being an "outsider" and neurodivergent. I lived in ignorance of my sensory capabilities until recently, when research revealed them.

Synesthetes and non-synesthetes have more in common than you might think. Some scientists believe all people are born synesthetes. Infants have hyperconnected neurons that

cause these bonus sensations, as proven by researchers led
by Daphne and Charles Maurer at Canada's McGill University.[1] But when infants are about four months old, their
brains begin to prune the connections. A father of modern
synesthesia research, Dr. Richard Cytowic, says that synesthetes are just getting a "conscious bleed" of things going on in everyone's brain. And I agree with him. I believe
all people are synesthetes—some latent, some consciously
aware. When I refer to *synesthetes* throughout this book, I
mean those people who are aware of the process. Conscious
synesthetes remain hyperconnected, like infants; in the case
of those with MTS, especially so.

Though my sensorium—the parts of the brain concerned with the interpretation of sensory input, or the seat
of my senses—is enormous and complex, it is also contig-
uous with a memory palace. That's a memory strategy the
ancients recommended; the idea is to use familiar places to
remember information. For instance, items on a grocery list
are easier to recall if you think of them in specific locations:
a loaf of bread on the bottom of your front steps, sticks of
butter in the mailbox, a quart of milk in your foyer. However, as a synesthete, I can never leave the palace.

There, the alphabet and multiplication tables I memorized as a child are still displayed in associated colors, which
make them indelible to me. The letter *k* is teal; *r* is terracotta; *7* is shiny ebony. And there, the sound of the cello
will always be a glassine and wavy and mocha-hued form
hovering three feet above the floor; the calendar on the wall
will always be rose quartz when turned to February; white
dots on a black background ebbing and flowing in silence

(a test for synesthesia) will always sound like rushing water. Nearly everything I perceive has a related color or sound, which is a helpful memory aid.

Interestingly, the fact that two or more sensory associations act as a memory aid is being exploited by businesspeople in a technique called *multisensory marketing*. Writing on the topic for his LinkedIn newsletter, author Raja Rajamannar, the chief marketing and communications officer for health-care business at Mastercard, observed, "Marketers need to find a way to cut through the clutter and effectively reach the consumer. I found that multisensory marketing in general, and sonic branding in particular, is an effective solution." He also noted, "Expressing a sentiment unique to the brand through multiple doors of perception—be it through the eyes, ears, or even taste—distinguishes a given brand's character from its competition and establishes a firm multidimensional identity."

My associations, apart from their value as mnemonics, are limiting in some ways (the pairing of sensations—my *a* is always yellow and my *5* is always navy-blue, for example—remain fixed for life). These pairings can feel like minutiae, and if I focus on them too much, I feel like I'm wading through molasses. It's hard to read text forward, for example, when I'm focusing on the colors. Most synesthetes have to learn to dial it down by not focusing on it.

Yet some blessed shafts of light are pouring in to my palatial prison. Punching holes in the ceiling, so to speak, are positive consequences of synesthesia—for instance, its association with metaphorical thinking and creativity. Synesthesia is very liberating on this macro level. Creating comparisons and brainstorming are much easier for synesthetes than for other people. These experiences feel like quicksilver when compared to the other aspects of synesthesia.

Synesthetes have a talent for metaphors. I like to call it a *sea of similitude,* for the feeling is oceanic. I've discerned synesthesia through people's language choices. Synesthetes seldom choose clichéd or commonplace comparisons. Writing on a mass of people gathered in Derry, Northern Ireland, for the anniversary of the Bloody Sunday human rights march, I referred to the crowd as "one heaving shoulder as a city cried." I noticed synesthesia in the metaphors of Nobel Prize–winning novelist Orhan Pamuk and later confirmed with him that he is a synesthete. He names characters for colors and has beautiful descriptions like "the heartache of dusk."[2] And I noted such metaphors in the lyrics of Broadway legend Stephen Schwartz and many other songwriters.[3] Think of how Pharrell Williams sings about feeling "like a room without a roof" in his hit song "Happy."

Life would be colorful enough for me with my synesthesia alone, but as I mentioned, I also have functional tetrachromacy, or four types of cones for color perception. Cones are photoreceptor cells located at the very center of the retina. Most people have three types, and each type is sensitive to colors at different wavelengths. Short-wave cones are sensitive to colors with short wavelengths, such as purple and blue; medium-wave cones are sensitive to medium-wavelength colors, like yellow and green; and long-wave cones are sensitive to long-wavelength colors, like red and orange. They also help us to see fine detail. The result of having the extra type of cone is super-color vision. I may see as many as one hundred times more color than is "normal." Not all the colors I see are beautiful, except in the way that secrecy adds allure to things. I tend to find the extra colors in nature, not in manufactured goods made by people with just three cone classes.

That is not to say any color appears ordinary to me. They are multilayered and sumptuous, even the ones everyone can see. Years ago, I often went by the window where my tailor, beautiful Nilüfer from Afghanistan, kept her spools of thread. Every time, I had to pause. Her name means "lotus," and like that flower, the colors opened and called to me. There was the cobalt thread for police officers' and firefighters' uniforms; mauve for that mother-of-the-bride dress; verdigris for the scrubs worn by the nurses in the local hospital; hunter green for the Catholic schoolgirls' uniforms; onyx for the bankers and the lawyers. Not only did the colors weave the story of my neighborhood, but they were as vivid as the neon sign in her window. If I looked closely at the tapioca-colored spool, I saw flecks of blood orange. The cobalt one had veins of smoke in it; the magenta had bits of gold. There is no such thing as a spool of thread with a single color; my eyes see every fiber.

There is not enough research to say where synesthesia ends and tetrachromacy begins. However, I am one of three American women who have publicly identified themselves as tetrachromats, and two of us are also synesthetes.

My own experience as a super-sensor makes me consider how everyone uses the senses and how much more could be perceived if people understood how powerful they are. Once my own sensing abilities were validated by science, I felt them start to expand—it was as if I'd been given permission to sense more widely. What you nurture will, eventually, flower.

Educators of young students are addressing the senses more nowadays, but many adults missed out on this when they were in school. What can adults do to compensate for this, and what more can teachers do to enrich young people's educations?

Anna Harris, an associate professor in the Department of Society Studies at Maastricht University, the Netherlands, wrote that everyone's senses—not just those people who've gotten an extra helping of genes—can be enhanced. In her book *A Sensory Education,* she referred to this as an opportunity for ongoing cultivation.

If you think of sensing as a skill rather than as a genetic ability, you can change how you approach it. Harris points out in her book that the French word meaning "cultivate," *colere,* brings together the ideas of both trying and skill, because *colere* means "to inhabit." "Sensory education means to inhabit sensing in social, bodily, and material ways, through trying to learn skills and trying to teach them," she wrote.[4]

Educator Maria Montessori's schools are a fine example of how to teach sensory education. Montessori said, "We cannot create observers by saying 'observe,' but by giving them the power and the means for this observation and these means are procured through education of the senses."[5] The Montessori method addresses nine senses. In addition to the standard five (touch, sound, taste, sight, and smell), there's *baric* (differentiation based on weight and/or pressure), *thermic* (the ability to sense various temperatures), *stereognostic* (the ability to distinguish an object without seeing it, hearing it, or smelling it, relying only on touch and muscle memory), and *muscular* (or kinesthetic—the ability to coordinate movement and balance).

Children as young as three still walk a white line taped on the floor to learn a sense of balance in the schools that bear her name. Objects of similar sizes but different weights are placed in tiny hands so students learn how to measure things in greater detail. Perhaps most effective of all, the children sometimes sit in complete silence. Instead of chatter, they hear the sounds of birdsong outside their classrooms. Ticking clocks are noticed for the first time. Hearing grows subtler this way, and self-control is also enhanced.

In *Dr. Montessori's Own Handbook,* Montessori wrote that she considered Helen Keller the best example of a successful sensory education. She was inspired by Keller and her teacher, Anne Sullivan Macy, as she developed her own approach to the classroom.

"In fact, Helen Keller is a marvelous example of the phenomenon common to all human beings: the possibility of the liberation of the imprisoned spirit of man by the education of the senses," she wrote. "If Helen Keller attained through exquisite natural gifts to an elevated conception of the world, who better than she proves that in the inmost self of man lies the spirit ready to reveal itself?"[6]

Keller herself often spoke on the senses. In her essay "Three Days to See,"[7] she imagined what it would be like to again have the gifts of sight and hearing she'd lost as a baby. "I who am blind can give one hint to those who see: Use your eyes as if tomorrow you would be stricken blind," she wrote. She went on: "Hear the music of voices, the song of a bird, the mighty strains of an orchestra, as if you would be stricken deaf tomorrow. Touch each object as if tomorrow your tactile sense would fail. Smell the perfume of flowers, taste with relish each morsel, as if tomorrow you could never smell and taste again. Make the most of every sense; glory in the beauty which the world in all the facets of pleasure reveals to you through the several means of contact which nature provides."

Rudolf Steiner, Austrian philosopher and founder of the Waldorf Schools, also emphasized the senses in his concept of education. His schools still work to develop them.

He enumerated twelve senses and put them in three categories. In the first category are the senses for understanding oneself: the senses of life, movement, balance, and touch. In the second category are the senses for perceiving the environment: smell, taste, sight, and temperature. The senses of knowledge make up the third category: hearing, speech, thought, and ego, with the latter two giving the experiencer a sense of what other people are thinking.

Steiner gave a number of lectures on the senses, including a talk titled "Man as a Being of Sense and Perception" in July 1921.[8] "I said a long time ago, and I am always repeating it," Steiner began, "that orthodox science takes into consideration only those senses for which obvious organs exist, such as the organs of sight, hearing, and so on. This way of looking at the matter is not satisfactory, because the province of sight, for example, is strictly delimited within the total range of our experiences, and so, equally, is, let us say, the perception of the ego of another man, or the perception of the meaning of words."

There were no brain scans that could indicate a person had synesthesia until the 1980s, and there was no supporting mirror-touch synesthesia research until the 1990s. Tetra-chromacy research is even more recent—the first tetrachro-mat was identified in England in 2010.

When I finally realized what was going on, I started to write about my experiences and those of others to raise awareness of these traits. And I have come to the conclusion that synesthetes should be educated differently from the start. I receive many emails from parents of young synesthetes who are struggling in school despite being very bright. This is very sad, especially when you consider the many synesthetes who have been extremely successful. There are many talented musicians and composers who are synesthetes, such as Chris Martin of Coldplay and the phenomenal Billie Eilish. Though not every synesthete is an artist—some are scientists, such as the great physicist Richard Feynman and the inventor Nikola Tesla—I believe arts education would draw out many of the abilities in this "neuro tribe." How can we help these kids cultivate their abilities? Perhaps their heightened sensitivity—many of them are empathic—causes them to be undone by their gifts when they're not in the right setting. A model could be developed to address the challenges they face and enhance the positive aspects of their traits.

One email I received from a parent stands out in my mind. A mom in the Pacific Northwest told me that her preteen synesthete daughter becomes physically ill in school during lessons about war and genocide. Her teacher and the other students did not understand the girl's physical reaction, and as a result, she was having issues academically and socially. With her mom's permission, I wrote directly to the girl and told her it was a superpower and not a flaw, and I recommended some books on being empathic (such as those by Dr. Judith Orloff, including *The Empath's Survival Guide: Life Strategies for Sensitive People*). I wish we could identify more such young people and encourage greater refinement of this sensitivity. There are many adult synesthetes in the helping and medical realms, and such children will be wonderful caregivers in the future if they are not shut down but encouraged.

Educators who work with autistic people have made a lot of progress in addressing the sensory needs of their students, and they provide wonderful examples of what can be done for young synesthetes. Reporting in the *American Journal of Occupational Therapy* in 1999, Jane Case-Smith and Teresa Bryan wrote about how their techniques benefited five preschool students with autism.

The children were observed playing and videotaped over a three-week period. After the baseline of their behaviors was established, an occupational therapist worked with each child individually and consulted with their teachers over the course of ten weeks. The researchers found that four of the five children became more engaged, and three had positive effects in mastery of goal-directed play.

"The results support descriptions in the literature

regarding the behavioral changes that children with autism can make when participating in intervention using a sensory integration approach," they concluded.

According to the American Occupational Therapy Association, many adults have undiagnosed sensory disorders, which can be an enormous challenge to their professional and social lives. We are living in a time when the awareness of the prevalence of sensory disorders is finally coming to the fore. Even people with "normal" sensory systems can benefit from monitoring sensory health. Occupational therapist Anna Jean Ayres pioneered sensory-integration theory in the 1970s, and her work still influences the field today. Her book *Sensory Integration and the Child*[9] has helped many parents and professionals who serve children with special needs, and I believe it should be a handbook for every adult. *Integration* means how we interpret the information the senses relay to us—how we perceive. Sensory processing makes up 80 percent of the nervous system, an enormous amount of biological real estate.

"Sensations are 'food' or nourishment for the nervous system," Ayres wrote. "Every muscle, joint, vital organ, bit of skin, and sense organ in the head sends sensory inputs to the brain. Every sensation is a form of information. The nervous system uses this information to produce responses that adapt the body and mind to that information. Without a good supply of many kinds of sensations, the nervous system cannot develop adequately. The brain needs a continuous variety of sensory nourishment to develop and then to function."

Ayres explained that sensations are streams of electrical impulses. "These impulses must be integrated to give them meaning. Integration is what turns sensation into perception. We perceive our body, other people, and objects because our brain has integrated the sensory impulses into meaningful forms and relationships." When you examine an orange, for instance, your eyes send information to the brain on its color and shape; your fingers send information on its textures; your nose sends information on its smell. All of these impressions are integrated in the brain to form your perception of the thing—an orange.

I asked an online group of adult synesthetes if they felt they would have benefited from a different educational model. I was asking for all of us, but queried synesthetes because of their sensitivity and ability to speak consciously of brain processes. What would such a curriculum look like? I wondered.

"I think it would be helpful for math in particular," responded Sonya Minnis of Brooklyn. "I could never do any kind of advanced math. I've mentioned in this group before how my numbers have color, gender, and personality. I could not disassociate those things while trying to solve math problems. So, I got many answers wrong because they felt or looked wrong."

Alison Wagstaff of the United Kingdom also struggled with math. "I was fantastic at maths until algebra, then fell apart! With hindsight, this must have at least in part have been caused by my synesthesia as I had clear colours for numbers and letters, but nothing after that. . . . Of course I only realized that synesthesia existed about ten years ago! I was also absolutely hopeless at science and languages—I couldn't cope with the boring monotone textbooks."

"School was horrible for me," Kaspar Anderegg of Berne, Germany, wrote. "It was extremely underwhelming intellectually, just 'waiting time' and unmotivating. And it was overwhelming because of the daily sensory overload, so much irrelevant information and distraction! I am sure we would have many real successful and happy synesthetes if there were special schooling for them."

Anderegg said that he'd focus on utility: "Some synesthesias are good for art, some for therapy; all have a very specific, but very efficient use. A musician could add music to colour synesthesia to his curriculum."

He added that a synesthesia education would be challenging because it would have to be very individualized. "It's impossible to tailor it to every synesthete, as we all perceive differently. The right environment for personal exploration, exchange, and advice from mentors would be important. And the teachers would have to know techniques to trigger the synesthesia in a positive and comforting way."

Aurélia McNicol of Switzerland participates in online forums like this because she wants to help her daughter, who is a synesthete. "My daughter could memorise the first one hundred digits of pi thanks to her synaesthesia much quicker and reliably than peers. I feel synaesthesia can be such a superpower if one is taught how to make benefit of it." This reminded me of the savant and synesthete Daniel Tammet, author of *Born on a Blue Day*. He used the associated colors of numbers to memorize pi to over 22,500 digits. While the additional colors he sees through synesthesia can be a challenge for many, some people, like Tammet, use it to their advantage.

Helen Wells of Victoria, Australia, believes educating young synesthetes about the hues that are so much a part of their world would help. "Possibly add an extra part to the curriculum about colour, as colour seems to be a bigger part of most synesthetes' world than the average person."

Jasmin Sinha of Germany said perhaps it's not the curriculum that should be changed but the way it is delivered. "I am a synaesthete with strong visualisation abilities—but also, with strong visualisation needs. These needs were by far not fulfilled in my school years [in Germany]. It was very boring because it all was too slow. No allowances were made for learning speed. This is something I hear from many synaesthetes. It would be worthwhile to make a link between synaesthesia and increased cognitive abilities / IQ."

A learning environment for synesthetes should not have too much sensory stimulation, she said, in order to prevent sensory overload.

Sinha practices something she calls "haptic self-feedback" in which she uses tactile stimulation to soothe herself into learning. "Up until this day, I remember the content of those school lessons where the teacher allowed me to doodle along. Knitting would also work."

She added that it was extremely hard for her to memorize auditory-only input. "But that's how our school system was (and, I believe, still is way too much). If only I would have known back then that I am a synaesthete, and how I can use my synaesthesia as a tool while studying."

I would like to see all people, not just synesthetes, evaluated and educated about their senses from an early age. While there are university programs being developed around the senses, very little is offered for children.

"The senses, being explorers of the world, open the way to knowledge," wrote Maria Montessori in *The Absorbent Mind.* "Our apparatus for educating the senses offers the child a key to guide his explorations of the world, they cast a light upon it which makes visible to him more things in greater detail than he could see in the dark, or uneducated state."[10]

And philosopher Immanuel Kant said in his *Critique of Pure Reason*, "That all our knowledge begins with experience there can be no doubt. For how should the faculty of knowledge be called into activity, if not by objects which affect our senses and which, on the one hand, produce representations by themselves or on the other, rouse the activity of our understanding to compare, connect, or separate them. . . . It is therefore a question which deserves at least closer investigation and cannot be disposed of at first sight: Whether there is any knowledge independent of all experience and even of all impressions of the senses."[11]

Before Helen Keller's other senses were opened by Anne Sullivan, Helen was a wild and belligerent child, frustrated by her limitations. Such was her rage in those early years that she tipped over the cradle of her younger sister. Think of how backward our sensory lives are now: our world is violent, and we have ruined our environment. In our sensory development, we are like young Helen, tipping over cradles.

How fantastic to learn that we are capable of so much more.

4

Taste Making

According to American linguist Carl Darling Buck, of the five senses, taste is the one most closely associated with sophisticated living. "Hence the familiar secondary uses of words for 'taste, good taste' with reference to aesthetic appreciation."[1]

Wine expert Jaime Smith embodies both definitions of the term *good taste*: he has a superhuman palate and a very refined demeanor. This is not a coincidence, as you'll see. But as a child, he was often told by adults that he had an overactive imagination and was too fussy and sensitive.[2]

Those characteristics were not weaknesses; they were abilities. Viewed in a more enlightened way, he was discerning and visionary from a young age. And now he has ascended to the heights of his profession. He has twice been named Best Sommelier in America by *Food and Wine* magazine.[3]

Dr. Marcello Spinella reported in the *International Journal of Neuroscience* in 2002 that olfaction (which is intrinsically related to the ability to taste) "is a sense that has close relationships with the limbic system and emotion. Empathy is a vicarious feeling of others' emotional states."[4] Olfaction and empathy are processed by the same neuro-anatomical structures, including the orbitofrontal cortex, the mediodorsal thalamus, and the amygdala, and travel along similar pathways. In fact, because of the brain's anatomy, smell is the only sense that goes right to the emotional part of the brain.

"We genuinely experience scent as the emotion we have attached to it," according to sensory educator Marzi Pecen. "Our hearts lift at the aroma that reminds us of a happy day at the beach, or our hearts break a little when we smell the aroma of a long-dead relative's soap."[5]

Just as progress is being made in the sensory realm, those who study psychology are also rethinking emotion. Being told one is "too sensitive" is now considered verbal abuse. Julie L. Hall, author of *The Narcissist in Your Life*, wrote in *Psychology Today*, "You may have spent years feeling confused and ashamed about why you're so touchy and easily wounded. Perhaps you believe you have good reason to feel upset but can't get out of the cycle of hurt and blame that seems to always leave you on the losing end of the argument."[6]

What does this have to do with the senses? Everything. There is no such thing as *too sensitive*—which is really a type of alertness and wakefulness—if you want to ascend to the realm of enhanced abilities.

Tea sommelier and sensory educator Marzi Pecen is a supertaster with a virtuoso palate. And, like Smith, she is a sensitive soul. "I love guiding people through the exploration of their senses," she told me. "You don't have to speak the language to enjoy tea or wine or any food, perfume, visual art, et cetera. You don't have to know the technical terms or intricacies to have it touch your heart."

Pecen had no road map for her abilities. "I had no outside influences when it came to sensory appreciation," she said. Because of her gift, she has a strong dislike of certain foods because she tastes them very intensely. When she was a child, her strong dislikes were not understood. "And it was legendary that I won the battle of 'You're not leaving the table till you eat one brussels sprout' when I was probably six or seven. I thought my dislikes were personal failings until probably the last decade or so. I never thought it was a side effect of a talent." Growing up, she developed an abiding fascination with tastes and scents, finding them inspiring and impossible to ignore. "It was almost nourishing, like another layer of beauty or experience."

Pecen pointed out that Americans can turn to other cultures to learn how to raise children into their senses. "We are typically not trained in recognizing scents. In fact, sensory education is almost nonexistent for children in North America."

While in a café in France, she was pleasantly surprised by the behavior of a little girl at the next table. In an essay she wrote, titled "The Scent of Tea," she described the scene:

"I marveled at a five-year-old in a Paris cafe as she rattled off what she smelled in her mother's wine glass, then did the same with a bite of cheese. She was encouraged to experience different smells and flavors in her world, probably from birth. This is how we all learn new flavors and scents."

It is not an accident that the little French girl was so observant. French parents and early-childhood educators work with children on such skills, according to Professor Marie-Anne Suizzo of the University of Texas at Austin. "In France, pleasure, or '*plaisir,*' is not a dirty word. It's not considered hedonistic to pursue pleasure. Perhaps a better translation of the word is 'enjoyment' or even 'delight.' Pleasure, in fact, takes the weight of a moral value, because according to the French, pleasure serves as a compass guiding people in their actions," she wrote on TheConversation.com.[7] "And parents begin teaching their children from very early childhood in a process called the education of taste, or '*l'éducation du gout.*'"

This home education is reinforced in the French classroom. In day-care centers, she noted, "even two-year-olds are served formal, yet relaxed, four-course lunches with an appetizer, main course, cheese plate and dessert."

Pharrell Williams told me in an interview that he believes

children should be educated regarding the senses. "The truth is some kids pick things up auditorily quicker, some kids pick things up visually, some kids pick things up kinesthetically and there's also olfactory and gustatory. And some synesthetes know more about something by the way it smells or tastes—like there are people who know what purple tastes like and it's not grape."[8]

Education is key, Pecen agreed. Some people claim they have a poor sense of smell or an impoverished palate for tasting or identifying flavors, but "they may just lack exposure and conscious identification of the fragrances and flavors," she said. "For example, if you have never tasted a cherimoya [a Central American fruit], you couldn't identify its flavor in a tea."

Pecen believes emotional sensitivity is the extreme of being attentive. "There is a calm and focus or attentiveness that has to occur to really experience what's going on," Pecen explained. "*Mindfulness* is a word that is overused and misunderstood these days. But in my understanding as a Buddhist practitioner, it is being present."

Smith agreed. "They go hand in hand. I don't think a lot of people are paying attention to their external world, hardly at all, and are more dialed into looking at pictures," he said. "I've been reading studies about people who take pictures of everything, where you no longer have that beautiful recall memory, because you're just looking at photos."

How did these two extraordinary people find their supersensory paths in a world only beginning to understand and respect their value? What can their journeys teach us about finding our own way?

Jaime Smith, now in his early fifties, works as a wine steward in Washington, D.C. He entered the restaurant business as a teenager, and, fatefully, he was allowed to go to staff tastings held by wine representatives even though he was underage. He told me that he could easily discern flavors and textures older staffers couldn't put a name to despite their experience. "I had zero wine knowledge at this time, but I had a dialed-in tasting skill that was natural. They were constantly amazed that this young, know-nothing kid was outshining them. Within a year, I was the one who was meeting with reps and buying wine for a very famous restaurant in Manhattan before I could legally drink. My first job led me to the wonderful life of wine that I have lived."

Even though Smith had to deal with people's ignorance, he asserted himself professionally and believed in his ability. This is the "chutzpah factor," a term coined by Kenneth R. Pelletier, an expert on biofeedback, psychosomatic medicine, and altered states of consciousness.

Dr. Pelletier kindly sent me a copy of his classic paper "The Chutzpah Factor in Altered States of Consciousness," published in 1977 in the *Journal of Transpersonal Psychology*.[9] In it, he presents examples of exceptional human performances, such as fire walking without getting burned, inserting a knitting needle through one's biceps without bleeding, and raising the temperature of one's hand as much as twenty-four degrees with the mind alone. All these performers had one thing in common—the belief that they could succeed in the task.

"Most individuals avoid situations where failure may lead to physical or psychological injury," Dr. Pelletier wrote with his coauthor, Dr. Erik Peper. However, these adepts "create situations in which the expectation of the people around them is sufficiently high that they are able to overcome their own fears and reservations."

Sommelier Smith overcame conditioning that undermined his self-esteem for much of his youth by surrounding himself with the best in his industry and pushing himself to excel.

Pecen's fascination with taste and smell gave her the chutzpah to enroll in Cinquième Sens, an elite olfactory institute in Paris. There, she studied perfumery. "It was the perfect place to be immersed in fragrance and education. I spent the summer smelling everything I could and drinking tea. I chose my apartment by its proximity to Mariage Frères Teas original tearoom in the Marais."

Pecen participated in their monthly tea-education sessions and helped with outreach courses for the tea shop. "I started incorporating tea tasting into mindfulness meditation workshops I was asked to lead. Eventually, I got certified as a tea specialist and became an instructor. As you can see, my sensory abilities were already in use, but I had no idea my experience was so different from other people's. Now I'm convinced that everybody has their own unique holographic experience of the sensory world."

Each person's tongue is unique—like a fingerprint—and researchers are working to develop biometric identification from them in a process similar to retina scans. The average tongue is about three inches long and has anywhere from two thousand to four thousand taste buds. Like all the senses, taste operates on a fine and minuscule level. Biting into a hot chili pepper may feel like getting an enormous wallop, but in fact, the process is beyond microscopic.

So how does taste work? Let's start with the tongue. It's covered with multiple small bumps, the taste papillae. Each papilla has clusters of taste buds, and each taste bud contains a number of taste cells. The taste bud looks a bit like orange sections arranged around a central area. The top part of each bud has a small indentation filled with fluid, and the chemical substances responsible for the taste of whatever you're eating are washed into this hollow for the sensory cells to analyze. The taste-producing substances activate the cell by changing specific proteins in its wall. The sensory cell then transmits messenger substances that activate additional nerve cells. These nerve cells pass information for a particular perception of flavor on to the brain.

Even the five basic categories of taste—sweet, salty, savory (umami), sour, and bitter—are being reconsidered. Unless you are a gourmet chef or sommelier, the sense of taste is often believed to be as simple as detecting those tastes. But it's far more complex and wondrous than that. Smith explained there are other categories of taste—for instance, kokumi.

According to an article in the *Journal of Biological Chemistry*, there are compounds that enhance sweet, salty, and umami tastes but have no taste themselves; these are called *kokumi substances*: "Substances with kokumi taste, which is distinct from the five basic tastes, have been used for many years in Japanese cuisine. GSH (gamma-glutamylcysteinylglycine), a typical kokumi taste substance, is abundantly present in food-grade yeast extract, which is commercially available and has been used to make foods taste savory and hearty."[10]

Carl Darling Buck wrote, "Hindus recognized six principal varieties of taste with sixty-three possible mixtures, while the Greeks believed there were eight."[11] Ones we don't highlight enough in modern times include pungent, astringent, rough or harsh, oily or greasy, and "winy."

Humans can discern at least one hundred thousand flavors. The tongue has receptors even for substances not traditionally believed to have flavor, like water.

In 2017, researchers discovered the human tongue likely has a set of taste buds that discern fresh water. "Ever since antiquity, philosophers have claimed that water has no flavor. Even Aristotle referred to it as 'tasteless' around 330 B.C.E.," Emily Underwood reported in the journal *Science.*[12] "But insects and amphibians have water-sensing nerve cells, and there is growing evidence of similar cells in mammals," wrote Patricia Di Lorenzo, a behavioral neuroscientist at the State University of New York in Binghamton. A few recent brain scan studies also suggest that a region of human cortex responds specifically to water, she reported.

This vast flavor bank evolved because humans need to be able to detect spoiled or dangerous food to survive. Scientists at the University of Buffalo in New York recently set out to find out more about how this works. And they were surprised to find redundancies in the system—backups, if you will, in the form of secondary proteins that detect harmful substances.

The need for safety and survival is probably how super-tasters evolved. "The supertaster gene could be a remnant of our evolutionary past, acting as a safety mechanism to stop us eating unsafe foods and toxins," the BBC reported.[13] In addition, supertasters often avoid sugary and fatty foods.

Neurodiversity among human beings is nature's way of ensuring the survival of the species. I believe author Steve Silberman said it best in a 2013 article for *Wired*: "In forests and tide pools, the value of biological diversity is resilience: the ability to withstand shifting conditions and resist attacks from predators. In a world changing faster than ever, honor-ing and nurturing neurodiversity is civilization's best chance to thrive in an uncertain future."[14]

"We've known for a long time that people don't all live in the same taste world," Professor John Hayes at the Penn State College of Agricultural Sciences told CNN recently.[15]

"There are supertasters and non-tasters," Hayes added. "Supertasters live in a neon taste world—everything is bright and vibrant. For non-tasters, everything is pastel. Nothing is ever really intense. Supertasters have twice as many taste buds as normal people," he said. "And they perceive bitter tastes more strongly and tend to crave salt, which blocks bitterness." Contrary to popular belief, these people avoid rich tastes in favor of bland ones. They also tend to avoid alcohol and cigarettes because of the strong flavors.

"Being called a supertaster sounds very fancy, but all that means is that I taste bitter more intensely than most other people do," explained Pecen. A doctor friend in sensory science asked her to answer a series of questions to determine if she was a non-taster, a taster, or a sensitive taster. "I fall in the category of sensitive taster," Pecen said. "Again, it seems like it would be fun, and it is, but not all the time. It means that I have some definite likes and dislikes regarding food, drink, and even clothing. I used to think it was a shortcoming not to like foods that others loved. I've repeatedly tried to like them, but my experience never changes. Now I accept that I will never like them and move on."

It's not a matter of preference, she said. Some of the foods most people enjoy taste completely vile to her. For example, she understands that for most people, beets taste sweet and earthy. "That sounds pleasant enough, but to me, they taste gag-inducingly bitter, sour, and like what I imagine the bottom of a dumpster would taste like."

For Smith and Pecen, the sense of taste happens not just in their mouths but in the space around them. Smith said that for him, wine undulates in the ether, not just in the glass, due to the bonus senses of synesthesia. A champagne or a German Riesling appears as long ribbons of yellow, green, and orange. The colorful forms it releases inspire him to call it "liquid art."

"For me, the most beautiful thing is smelling [wine]," Smith said. "Because the smell, when I come into this gorgeous alcohol perfume, first sets off my synesthesia into just wonderful, billowing clouds of smells, and colors, and textures."

That first smell sets off subchambers. "Everything starts clicking into boxes. I'm like a safecracker because all these smells and all these things start just clicking away and building their own little shape. And suddenly, I have a nebulous glob of information that is somehow only programmed into my brain. And then I go into drinking, and then the drinking supersedes and overemphasizes what I had gotten through smelling. But it's literally so small of an impression compared to what the smell is. It's really more just a visceral commentary about the smell."

A polysynesthete, Pecen also strongly associates colors and textures and shapes with aromas and tastes. But her strongest impressions are outside the olfactory world. "The one I relate to most is probably that every movement or step in ballet has a texture and color. I didn't know that was not how everyone experienced dance or movement. I have it less as part of everyday life, so it may relate to music as well."

In the musical synesthesia library in Pecen's elegant mind, a favorite piece, Debussy's "Clair de Lune," is "a pale bluish-green chiffon-thin, soft silk fabric."

If you are not a conscious synesthete, you can still create secondary associations for yourself to make scents and flavors more memorable and richer. I recommend assigning shapes to flavors; for example, salt could be a triangle. Tying a sound to a taste would also help. For instance, peppermint could be the sound of a bell.

How indelible can these things be? They can last a lifetime. Pecen related what one whiff did for her.

She was participating in a group event at a fragrance

shop in New York City and sniffing a line of fragrant teas and cocktails. "They were perfume snapshots—frozen moments in time of the creator. We used ones related to a memory of gardening, one of going to the circus, another that was spending the evening in an outdoor *onsen*, or hot spring, in Japan." Pecen found them delightful, but not evocative of her own memories. "Then I started smelling other scents in the line. The aroma of an old barrel that my father aged wine in: sour, smoky, floral, a bit spicy, and richly woody. Others were recognizable—a beach, a forest— but they had no personal association with me. Then one stopped my heart. It was called 'Old School Bench.'"

It transported her to a time decades earlier.

"I did not expect the aroma of the tenth floor of the Fine Arts Building in Chicago. The benches and doors, the aftershave the elevator operator wore, the rosin and sweat from the ballet studio, the faint patchouli and incense leaking from the penthouse yoga studio. The ever-present aroma of charcoal, pencils, paint, wax, and wood emanating from the artists' studios up the hallway and the slight whiff of lemon furniture polish from up the hall; the odd roselike aroma from the strange memorial chapel a few floors below."

Pecen also emphasized the importance of smell to taste. "The senses of taste and smell are closely linked. It is the sense of smell that differentiates the stimuli and combines with the mouth information to make what we call *flavor*. Flavor is created in our brains, not our mouths." It's the brain that makes us think that these two separate senses create one experience in the mouth, she explained. "The volatiles are sucked up into the nasal cavity and to the olfactory receptors by the vacuum created by swallowing. All the swishing and bubbling to aerosolize a liquid in our mouth is theater. That doesn't make us experience the aroma any more intensely. Or at least that is true for many tasters I have asked to experiment with the process and me."

Pecen said she'd like people of all abilities to think about what they're tasting. "We all have these events locked in us. Unlocking them can be a challenge or a joy. It takes attention to experience them, or they can drift away and be lost. The key to all the wonder these moments bring is attention. Being willing to be open to experience judgment-free with undivided attention."

She added that people are missing experiences because "we have so much information coming at us all day long; we filter so much out. We would be overwhelmed if we let it all in, but how enriching it is to be open to experience more. We gulp water and never experience the taste of it, nor the nuances of foods or flavors." Sommeliers don't have a better sense of taste than other people, but they pay attention and they practice, and they have a vocabulary to share the experience. But, she says, "you don't need to have the vocabulary or even be able to identify what you are experiencing." You just have to be open to the sensory stimulation, "the texture of a fabric held and how it changes when stroked, the temperature of a beverage and how it evolves in your mouth."

Being positive aids in opening the senses. Jaime Smith and Marzi Pecen reflected on that big idea, demonstrated on the tiniest scale.

"For you to smell something, an actual molecule of that

item has to find its way to your olfactory epithelium up in the roof of your nasal cavity," Pecen said. "That structure is a part of your nervous system—uncovered, unprotected. An exposed bundle of nerves—an extension of the brain. A molecule touches your brain, then is perceived as an aroma. It just takes one molecule. The aroma of lilacs is not wafting from the bush in cartoon odor waves. It's a molecule that has cut loose from a fragrant cell, traveled through the atmosphere, and nestled into a tiny spot in your body. It is *in* you. You are experiencing it directly."

"I like to think that I am an instrument tuning in to the wonders of the cosmos and that my synesthesia is my galactic harmonic," said Smith. "For me, art is the highest human achievement, and I like to try and express it in everything I do."

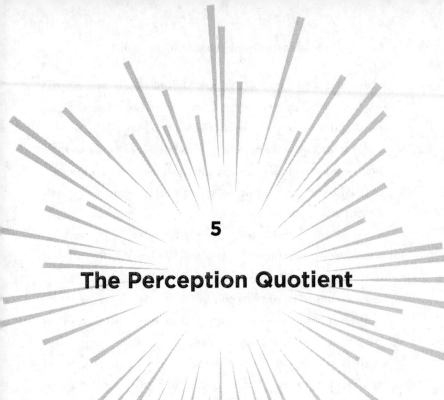

5

The Perception Quotient

Given how powerful the senses are, it's possible that PQ (the perception quotient) is even more important to survival than one's IQ or EQ (emotional intelligence quotient).

In fact, in 1884 when Sir Francis Galton created the first test for intelligence, he examined his subjects' senses. He checked vision, hearing, and reaction times to various stimuli.

A few years later, in 1890, James McKeen Cattell pioneered the first standardized mental test. He also measured the speed and accuracy of people's sensory perceptions.

Measures of sensory perceptions fell out of favor because they could not predict academic achievement. Mensa, the high-IQ society, stated on its website that "they are probably imperfect measures of anything we would call intelligence." I respectfully and strongly disagree. A person's IQ is not the sole predictor of success. We have come to learn that emotional intelligence, social intelligence, and other indexes should be used in combination with IQ tests.

Intelligence tests measure a person's mental abilities and compare them with the abilities of other people. But there is not actually wide agreement on what the word *intelligence* means. "For example, debate has surrounded whether intelligence should be considered an inherent cognitive capacity, an achieved level of performance, or a qualitative construct that cannot be measured," according to researcher Ellen B. Braaten.[1] Traditional intelligence tests measure specific cognitive abilities that can predict whether a child will do well in school, but they don't measure other skills, such as interpersonal abilities.

EQ, pioneered by psychologist and *New York Times* science writer Daniel Goleman, is a better predictor of overall success in life. He described five key components in his bestselling 1995 book, *Emotional Intelligence: Why It Can Matter More Than IQ*. They are "self-awareness, self-regulation, motivation, empathy, and social skills." EQ is a plastic ability, just like PQ.[2]

But what is success? The ultimate success is the ability to

survive in a world growing more fragile by the day. I believe PQ is a better fundamental measure of human potential— more primal and more essential. The early history of intelligence testing was rife with biases against women and minorities, and it stank of eugenics. I find it poetic and poignant that indigenous hunter-gatherers have higher perception quotients than people in more settled societies.

"Hunter-gatherers who live off the land in the forests of Malaysia have a far more evolved sense of smell than more settled people," noted the authors of a study published in the journal *Current Biology*.[3]

Professor Asifa Majid of Radboud University and her team had previously established that the Jahai people of Malaysia have a vast number of words for scents. "There has been a long-standing consensus that 'smell is the mute sense, the one without words,' and decades of research with English-speaking participants seemed to confirm this," said Professor Majid, who specializes in language research. "But the Jahai of the Malay Peninsula are much better at naming odors than their English-speaking peers. This, of course, raises the question of where this difference originates."

Researchers asked members of two Malaysian tribes— the Semaq Beri, who are hunter-gatherers, and the Semelai, who are somewhat settled horticulturalists—to name colors and smells. They found that the horticulturalists could name colors without a problem but struggled to name smells, while the hunter-gatherers, who spent more time on the land looking for food, could name both colors and smells easily.

These findings confirmed that a hunter-gatherer lifestyle

enhances sensory ability, which contradicts the theory that it is the structure of the brain alone that determines the sense of smell. "For the hunter-gatherer Semaq Beri, odor naming was as easy as color naming, suggesting that hunter-gatherer olfactory cognition is special," the scientists wrote.

So what hope do those of us who live in industrialized societies have of moving in a more advanced sensory direction? It will depend on what evolution demands.

Anthropologists used to think that human evolution had slowed down. In fact, some scientists believed humans had completely stopped evolving. Geneticist Steven Jones of University College, London, told *Nova* that while natural selection is still a factor in parts of the developing world, particularly in Africa, where people constantly struggle with lethal diseases, in the developed world, human evolution is "genuinely *over*."[4]

But in July 2008, Professor John Hawks, a paleoanthropologist from the University of Wisconsin, reported that human evolution has accelerated a hundredfold in the past five to ten thousand years.

"People who lived ten thousand years ago were much more like Neanderthals than we are like those people," Hawks told *U.S. News and World Report*.[5] "We've changed." He came to this conclusion after comparing chunks of DNA from 269 people from around the world. "Over time, DNA accumulates random mutations, just as the front of a white T-shirt tends to accumulate spots. The bigger the chunks of DNA without random spots, the more recently it had been minted."

Hawks figured out that recent genetic changes accounted

tor about 7 percent of the human genome. The rapid growth of the world's population is key to this, he said. An increase in people means an increase in mutations.

One of the major changes in the genome happened around ten thousand years ago, when the ability to digest lactose became common among people in the West. Today, more than 70 percent of Europeans can drink milk with no problem. Not surprisingly, this change coincides with the domestication of farm animals. In addition, humans once needed wisdom teeth to tear apart roots and other raw foods, but today, 35 percent of people are born without them. And every person with blue eyes today descended from a single human who was born near the Black Sea only ten thousand years ago.

While it might seem that in a technologically advanced world, evolving to be more conventionally intelligent would be advantageous, Bill Nye the Science Guy presciently warned us long before the COVID pandemic that it is resilience to disease that will win the day. "More likely than a future race of hyper-smart people who out-compete the rest of us is a strain of *Homo sapiens* that can beat a disease," he wrote in *Popular Science* in 2014.[6] "Probably the most important evolutionary sieve that any future person is going to have to get through is going to have to do with germs and parasites."

I would argue that not only resilience to disease but avoidance of it in the first place is the key to survival, as we saw during the COVID pandemic. And that's where a highly developed sensory system comes in.

In the *Proceedings of the National Academy of Sciences*

in 2017, scientists from the Karolinska Institute in Sweden wrote, "The human brain is much better than previously thought at discovering and avoiding disease. Our sense of vision and smell alone are enough to make us aware that someone has a disease even before it breaks out. And not only aware—but we also act upon the information and avoid sick people."[7]

In the endless race between organisms and pathogens, the human immune system has evolved to lower the risk of infections, principal investigator Professor Mats Olsson observed. "The present study shows that we can detect both facial and olfactory cues of sickness in others just hours after experimental activation of their immune system."

In this study, his team injected half of a group of volunteers with a compound known to cause a robust immune response (leading to symptoms like fever and fatigue) and the other half with a placebo. Several weeks later, the researchers repeated the process, but this time, the individuals received whatever injection they hadn't gotten previously. The scientists took body-odor samples and photographs of the volunteers after each round of injections.

A separate group of participants were then exposed to the smells and images of the volunteers, and while their brain functions were being monitored, they were asked to rate how much they liked each person on the basis of a photograph and an odor. Next, the participants were asked to look at the photos and rate each person in terms of attractiveness, health, and social desirability. Then they were given the odors to smell and asked to rate each one for intensity, pleasantness, and health.

"Faces were less socially desirable when sick, and sick body odors tended to lower liking of the faces," reported the scientists. The brain combines all the signals and makes a judgment about the health of the other person.

If people were brought up to have high PQs, not only would they have better survival skills, they would have more access to and more enjoyment of the senses. These people of the future would have a sense of being here and present and aware.

American poet, essayist, and naturalist Diane Ackerman is that kind of evolved human being, someone in full awareness of her senses. In her bestselling book *A Natural History of the Senses,* she joyfully and lyrically soars through a celebration of all the senses. As an homage, a crocodilian pheromone molecule—dianeackerone—was named after her. Here is one of my favorite passages from her crystal-clear doors of perception:

> I touch the soft petal of a red rose called "Mr. Lincoln," and my receptors translate that mechanical touch into electrical impulses that the brain reads as soft, supple, thin, curled, dewy, velvety: rose petal–like. When Walt Whitman said: "I sing the body electric," he didn't know how prescient he was. The body does indeed sing with electricity, which the mind deftly analyzes and considers. So, to some extent, reality is an agreed-upon fiction. How silly, then, that philosophers should quarrel about appearance and reality.[8]

Another person with high PQ is David Abram, an ecologist and philosopher whose work has had a profound effect on the environmental movement. He wrote in *The Spell of the Sensuous* that "the sensing body is not a programmed machine but an active and open form, continually improvising its relation to things and to the world." He and Ackerman are advocates of both the natural world and the senses.

After discussing some of the damage humans are doing to the planet, Abram wrote, "These remarkable and disturbing occurrences, all readily traceable to the ongoing activity of 'civilized' humankind, did indeed suggest the possibility that there was a perceptual problem in my culture, that modern, 'civilized' humanity simply did not perceive surrounding nature in a clear manner, if we have even been perceiving it at all."[9]

Those who spend more time in nature have a heightened sensory awareness. But in a 1989 report to the U.S. Congress, the Environmental Protection Agency noted that Americans spend a shocking 90 percent of their time indoors.[10] Other studies showed this is true in Canada and Europe as well. In 2015, Richard Corsi, now the dean of engineering and computer science at Portland State University, noted that the average American has a life expectancy of seventy-nine years, and seventy of those years are spent inside buildings, "a greater percentage of time than whales spend submerged below the surface of the ocean."[11] This dulls our senses and makes us ignorant of environmental crises.

✳

Raising everyone's PQ would also counter one of society's biggest problems: desensitization.

"Youth are exposed to large amounts of violence in real life and media, which may lead to desensitization," Sylvie Mrug of the University of Alabama at Birmingham reported in the *Journal of Youth and Adolescence* in 2014.[12] She and her colleagues examined the associations between exposure to violence (both in real life and in the movies) and empathy (among other variables). They found that subjects who had experienced high levels of real-life violence had reduced empathy, and, in males, "higher levels of exposure to real-life violence were associated with diminished emotional reactivity to violent videos. Thus, youth exposed to higher levels of real-life violence do show some signs of emotional desensitization involving lower empathy."

But we can regain our sensitivity and humanity. Often, art leads the way.

Artist and professor Carrie C. Firman suffers from fibromyalgia and has synesthetic color impressions for her pain. She made her discomfort visible to others by creating a trench coat fitted with different-colored LED lights that glow in the places where she hurts. *Sympathy Pains*, the name of the art piece, features metal cones, elastic straps, and bags of rice of different weights. Pressure sensors vary the intensity of the lights to show the level of pain associated with movement.

"I made this piece during the summer in the middle of my MFA program, while in a wheelchair recovering from a broken leg," Firman told me recently.[13] "What was so revealing to me that summer was that as soon as my pain was visible—a toe-to-knee cast and the wheelchair—

people cared without my request. What if my chronic pain, affecting almost my entire being, was visible? It was clear that words failed when describing it, and thus the coat was born."

Firman, a tenured professor of graphic design at Edgewood College in Madison, Wisconsin, shared her wisdom about different forms of intelligence. "A whole other route here could be Rudolf Steiner Waldorf Schools or Montessori models for learning versus our typical Western academic system of knowledge acquisition," she said. "Studies repeatedly show that experiential and interactive learning is a far better model." Creating ties between subjects rather than separating them would also help.

Firman noted that Sir Francis Galton, who created the first test for intelligence, also coined the term *synesthesia*. "This was also during that first era of the spike of studies on synesthesia. No coincidence. We think of the Victorians as so tightly bound, but art was very alive at the time, and perhaps because we knew *less* about the mind, there was more freedom of research in this area. Let's blame Freud for twentieth-century restrictions. We can always blame Freud. Ugh."

Anthropology professor David Howes, the codirector of the Concordia University Centre for Sensory Studies in Montreal, told me that the World War I exhibit in London's Imperial War Museums offered a way for art to reverse desensitization to the horrors of war. "You go in the trenches, and they simulate the experience, so it makes it real."[14] The permanent exhibit puts visitors into the battle and features a German tank and an airplane overhead. Howes said it made him feel the experiences of the soldiers.

In 1995, I visited the United Nations to see a land-mine exhibit. At the time, I was writing about Kenan Malkić, a boy who lost both of his arms and a leg when he stepped on a mine in Bosnia. I wanted to understand his experience, which was, and in many ways still is, unfathomable. I will never forget seeing the (now-deactivated) land mines—many of which had been intended for children—in sandboxes set throughout the lobby of the UN General Assembly building. Nor will I ever shake the experience of not knowing if a floor tile I stepped on would trigger the muted sound of an explosion. Bosnia—and the wider land-mine-riddled world—became real to my senses, not just something I watched in a movie or on the evening news. Like the World War I exhibit did for David Howes, the UN installation made me care deeply in the way the best art can.

During this time, I came to know Muhamed Sacirbey, who was Bosnia's ambassador to the United Nations during the war and that country's foreign minister. I'd known he was highly empathic—advocating for the victims of war beautifully and effectively—but I also learned that he experiences synesthesia. His wife, Susan Sacirbey, is also profoundly empathic. They spend a great deal of time in nature.

Sacirbey effectively used art to convey his country's situation, and he was instrumental in creating the Concert for Sarajevo with U2 and Luciano Pavarotti. He told me that he finds art valuable because "it enhances different ways to perceive."

When I asked him to describe how he experiences empathy, he said, "I do feel the pain of others, including animals, but I'm not certain to what degree, if any, it manifests as reflection in my own body. Rather, I feel obliged to do something to relieve such pain or fear, even if it necessitates some more overt action."[15]

As Carrie Firman noted, "We have colonized our own senses, creating a society based on assimilation, work, and rigid social structures." Westerners are reaping what they have sown.

When the world is safer for sensitivity, all of our senses will be heightened.

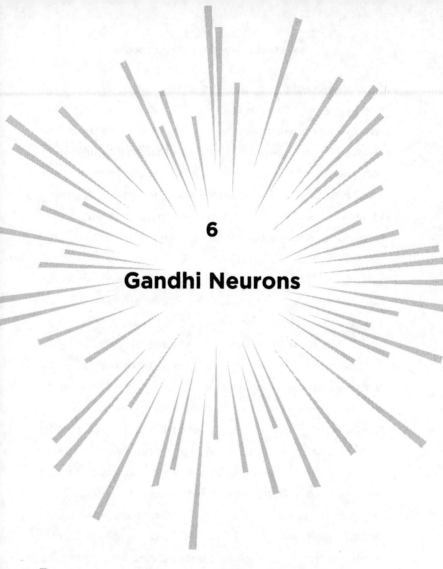

6

Gandhi Neurons

Patricia Lynne Duffy, a member of the United Nations Staff 1 Percent for Development Fund committee, was safely ensconced in her Manhattan office, but when she read the proposals before her, she felt the pain of the world's most fragile people.

Duffy's arms ached with the weight of the buckets of water that village girls in Ghana had to carry several miles every day to supply clean water to their families. Her heart was heavy with the despair of rural widows in Bangladesh who lacked status and autonomy. She felt physical pain when she read about children in Haiti who had been injured in an earthquake. Duffy has mirror-touch synesthesia—a sense of profound empathy for others resulting in physical responses in oneself—and she felt as if all these people's sufferings had traveled across oceans and mountain ranges and settled in her petite frame.

"In my mind, when I read project proposals, I cannot help but visualize and feel the descriptions I am reading. . . . I have a visceral reaction," she told me.[1]

Mirror-touch synesthesia may have its basis in brain cells called *mirror neurons*. They were first discovered in the brains of macaque monkeys in the early 1990s; Italian researchers observed that certain neurons in a monkey's brain fired when it picked up a peanut *and* when it watched another monkey pick up a peanut—that is, the neurons reflected—or mirrored—the behavior the monkey was watching. Humans seem to have a similar mirror-neuron system (although the research is more difficult to do in human subjects); mirror neurons have been found in several regions of the human brain: the premotor cortex, the supplementary motor area, the primary somatosensory cortex, and the inferior parietal cortex.[2]

Neuroscientist V. S. Ramachandran noted that the discovery of mirror neurons was the most underreported scientific finding of its decade. He predicted that mirror neurons would one day "do for psychology what DNA did for biology," that they would explain previously mysterious processes of consciousness. He calls these brain cells "Gandhi neurons," and he believes they are the very basis of human civilization. The hominid brain has been approximately the same size (with, possibly, the same capacity for intelligence) for the past 250,000 years, but the traits we associate with being human (wearing clothes, making art, and using tools, for example) didn't appear until about 40,000 years ago. Ramachandran suggests that the evolution of these mirror neurons is what drove humanity's great leap forward.

"Let's look at human evolution," Dr. Ramachandran said in a TED-Ed Talk in 2013.[3] "It turns out there's something very important that happened around seventy-five to a hundred thousand years ago, and that is there's a sudden emergence and rapid spread of a number of skills that are unique to human beings, like tool use, the use of fire, the use of shelters, and, of course, language and the ability to read somebody else's mind and interpret that person's behavior." He believes that it was the sudden emergence of a sophisticated mirror-neuron system that allowed people to emulate and imitate the behavior of others.

"When there was a discovery by just one member of a group, for example, to use fire or a particular type of tool," he said, "instead of dying out, the new skill spread rapidly and horizontally across humanity."

There are a few theories about how MTS works. The threshold theory posits that MTS synesthetes have hyperactivity in the mirror-neuron system that leads to heightened sensitivity in their bodies. The self-other theory suggests that MTS is associated with an impaired ability to distinguish oneself from others. A third theory, recently advanced by Shenbing Kuang in *Frontiers in Human Neuroscience,* combines both hypotheses.[4]

Everyone has mirror-touch neurons, but 1.6 percent of humanity have the associated synesthesia, meaning that they consciously experience other people's sensations. This brings us back to Patricia Lynne Duffy. Duffy trains diplomats and other staff at the United Nations, and she holds a master's degree from Columbia University Teachers College. "When I conduct training sessions, I always feel a strong sense of what my course participants are experiencing—where they are engaged and exhilarated, where they are confused, when they are defensive, resistant. I can feel their emotions 'mirrored' in my own internal space," she explained. "Those realities help me determine the direction of the session and how aspects of the material can be best presented and absorbed." She likens it to a virtual-reality experience but without any headsets or goggles.

Duffy said that a powerful virtual-reality film about a twelve-year-old Jordanian girl in a refugee camp gave viewers a chance to try on "empathy goggles." *Clouds Over Sidra,* made in partnership with the UN, inspired viewers to donate a billion more dollars than expected when it debuted at the Davos World Economic Forum.

Duffy notes, "In a sense, having MTS is a bit like having an internal virtual-reality mechanism and response." She reminded me of the words of philosopher David Hume: "No quality of human nature is more remarkable, both in itself and in its consequences than that propensity we have to sympathize with others. The minds of men are mirrors to one another."

I suggested to Duffy that it must be very hard to be a firsthand witness to the world's never-ending crises.

"Every day, when we hear or read the news, many of us are shocked by what often seems a stunning lack of empathy on the part of some world leaders," she said. "And this lack may emerge from old models of what it means to be a leader. Too often, being strong has meant being emotionless—or ignoring one's emotions—even engaging in ruthless behavior to reach the leader's goal."

But, she added, "These days, I see signs of hope in a changing attitude toward the importance of empathy in leadership. Many have realized that organizations and countries are composed of people with feelings and values that need to be acknowledged and respected. Recognizing and working with this reality is the way to make an organization, or a nation, sustainable."

The Center for Creative Leadership cites empathy as the number-one quality for good leadership, Duffy said. "Why? Because people are more likely to give their best when they feel they are understood and respected."[5]

When the feeling of being respected and understood is absent, people "drift away," she explained; they feel alienation, resentment, anger, or despair.

"The more societies as a whole learn to value empathy as not only an important quality but as an essential quality for a leader, and the more we learn about channeling and cultivating this sentiment, the more chance we have of realizing the dream of a peaceful world . . . the dream of some of the greatest empaths in history: Mahatma Gandhi, Nelson Mandela, Dr. Martin Luther King, Mother Teresa,

Malala Yousafzai, Greta Thunberg—and so many more who have moved the world forward."

Mirror-touch synesthesia can have positive effects, but it also has a downside. Extreme empathy can be overpowering at times, even debilitating; it can blur boundaries and cause distress.

Carolyn "CC" Hart, a writer and massage therapist in San Francisco, has MTS, and she has also had to learn to manage these extreme senses. It is very difficult for Hart to do bodywork on people in pain because she feels their discomfort in her own body. She gets shooting, stinging pain down her legs when she sees another person's injuries. She has been this way since childhood. Hart believes people with MTS need to engage in more self-care. "Sometimes I think we've got a lot of wonderful capacities, but damn it if we don't have to take care of ourselves maybe even a little bit more than most people. And that might mean some sensory deprivation."[6]

The healer described a small chain of spas in San Francisco called Reboot. It has floatation pods, and "you can do a pod with or without music and float into that abyss." Hart strips down and puts in earplugs. "And I get in there and I drop in and I float." Splashes of the water or the sensation of the water on her skin create color patterns in her visual field. "I'll float along and accidentally touch the side of the pod, and it's a gentle touch. A whole bunch of colorful patterns will pop up. So even in a sensory-deprivation environment, I still get a lot of intense sensory input because of synesthesia."

I didn't "come out" to people about my own MTS until

a couple of years ago, when award-winning science teacher Theresa Kutza invited me to speak on synesthesia at New Dorp High School, a public school in New York City. She had learned about my work when a librarian at the school, Gloria Medina, told her about my book *Struck by Genius* (about a man who develops synesthesia and savant syndrome following a head injury).[7]

I spoke to ninth graders in the school's science and math institute about synesthesia and savantism. Then one of them asked if I, too, experienced mirror-touch. "Yes," I said, a little reluctantly, having only recently come to terms with this, despite the fact that I'd experienced it all my life. That's when the hands started going up.

The students wanted to know if I could feel the specific aches and pains they were feeling. It was the first time anyone had ever asked me to do this, and I felt nervous, despite the signals from their bodies that were already poking me. I answered correctly six times out of six, reporting pains ranging from sprained toes to an achy back to a headache to hand pain (one young man surreptitiously stuck the point of a ballpoint pen into his palm to test me).

I looked over at the children's teacher, Theresa Kutza, and she was crying. I started to lose it a little too. It was so validating and such a huge emotional relief because I hadn't wanted to disappoint the students.

I realize that for mirror-touch people like me, caring about other people in pain is not a choice but a biological imperative. How can we not care about others' pain if we feel it in our own bodies? Perhaps learned empathy among people who don't experience MTS is more noble and the greater mystery.

Dr. Ramachandran said at the conclusion of his TED-Ed Talk that he wasn't speaking in the abstract or metaphorically when he referred to mirror-touch neurons as *Gandhi neurons* and *empathy neurons*. "You're quite literally connected by your neurons and whole chains of neurons around this room talking to each other. There is no real distinctiveness of your consciousness from somebody else's consciousness. This is not mumbo-jumbo philosophy—it emerges from our understanding of basic neuroscience."

He concluded: "I'm saying the mirror-neuron system is the interface allowing you to think about issues like consciousness, representation of self, what separates you from other human beings, what allows you to empathize with other human beings." Mirror neurons drive "the emergence of culture and civilization, which is unique to human beings."

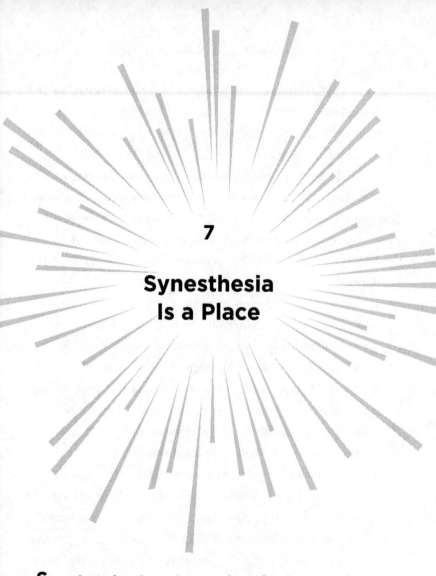

7

Synesthesia
Is a Place

Synesthesia has been incorrectly defined, in my humble opinion, as "crossed wires in the brain" or "mixed-up senses." In fact, synesthetes have the same primary response to a stimulus as neurotypical people do. If the numeral 5 appears in newsprint, I know that it is black on a white (well, somewhat beige) background. However, simultaneously, I see navy blue around and above that number, like an aura. Therefore, I've created what I believe is a much better definition: *Synesthesias* are traits in which a sensory stimulus yields the expected sensory response plus one or more additional sensory responses.[1]

Dr. David Eagleman, a Stanford University neuroscientist who studies brain plasticity, sensory substitution, and time perception, says we should think of synesthesia as being like two countries with porous borders. "In most brains," he told *The New Yorker,* "they stay separate. But, in the synesthete's brain, they communicate."[2]

I wondered if the two countries Eagleman was referring to were the subconscious and the conscious minds. One of the fathers of modern synesthesia research, Dr. Richard Cytowic, has long believed that we are all synesthetes; it's just that some of us get a "conscious bleed" and can speak of the subterranean workings of the brain. I emailed Dr. Eagleman hoping that he could help me understand this better.

He replied that what he'd meant was that the country of numbers, for example, has porous borders with the country of colors, that the bonus senses synesthetes have are due to open paths and high connectivity between the regions of the brain where these perceptions "live."

These neural-pathway neighbors metaphorically borrowing sugar from each other can result in various interconnections, like actor Tilda Swinton tasting cake whenever she says the word *table,* and how I associate the letter *A* with yellow and Wednesday with indigo. Synesthetes like Swinton would hear the word *table* just like the rest of us, but they get bonus impressions on top of the usual ones.

But what about that conscious-versus-subconscious aspect? I turned to Dr. Cytowic for an answer to this. Could synesthetes be accessing more of the subconscious mind than neurotypicals?

"Absolutely," he replied. "Way back at the beginning, I said that synesthetes were accessing a normal brain process that is typically unconscious in many of us. You would likely be sensitive to or subliminally aware of other stuff besides synesthesia."

After speaking with top neuroscientists, I realized how very wide open the synesthetic mind is; it may be a matter for the emerging science of consciousness because its experiences are so extraordinary.

In the journal *Psyche* in 1995, Dr. Cytowic wrote that we often think of the flow of neural impulses as linear.[3] "We think of perception as a one-way street, traveling from the outside world inwards, dispatching a linear stream of neural impulses from one relay to ever more complex ones, so that the process is metaphorically like a conveyor belt running through stations in a factory, until a perception rolls off the end as the finished product." But, he suggested, instead of looking at the finished product, why not focus on an earlier stage? After all, if you're looking for similarities between family members on a family tree, the relatives who are closest to the tree's trunk are going to resemble each other far more than relatives who are on the distant branches do. In terms of what a synesthete perceives, between the eye and the visual cortex, there are many possible areas for transformation.

"The image we see on the screen when watching television is the terminal stage of the broadcast," Dr. Cytowic said in an apt analogy. If you could intercept the transmission somewhere between the studio camera and the television, you would presumably see a different image from what others see when they look at the final product on the TV screen.

"We can similarly propose and test the concept of synesthesia as the premature display of a normal cognitive process. This implies that we are all synesthetic and that only a handful of people are consciously aware of the holistic nature of perception."

I would like to acquaint you with some of the many wild and gorgeous visions that appear when you intercept the signal before it hits the screen.

A filmmaker once asked me to describe synesthesia in a single word. "Place," I said without hesitating. The things many synesthetes see in response to stimuli appear in a sort of hyperspace. For example, the sound of the higher notes on a piano are cool, lemony glass columns near my right cheek. There is a sense of dimension and position to the experience. It feels very much like what UN instructor Patricia Lynne Duffy described in the previous chapter: virtual reality without the goggles.

In the *Psychonomic Bulletin & Review* in 2009, researchers reported finding evidence that humans can intuit four-dimensional space despite living in 3D.[4] "It is a long-lasting question whether human beings, who evolved in a physical world of three dimensions, are capable of overcoming this fundamental limitation to develop an intuitive understanding of four-dimensional space," they wrote.

The research into 4D resonated with Duffy, who, in addition to having mirror touch synesthesia, sees photisms in space: when she visualizes the alphabet, the colorful letters rise in the air from left to right.

"That research ... takes the mind to a new place," she commented. It reminded her of something the artist Max Beckmann said about the painting process: "To transform into two dimensions is for me an experience full of magic in which I glimpse for a moment that fourth dimension which my whole being is seeking."

"And yes," Duffy said, "it sometimes seems there can be such momentary 'glimpsing' in synesthesia's internal landscapes of language, music, and time."

In childhood, I discovered that the forms and colors I saw in the air around me could not be held, touched, or moved; they were both there and not there. Later in life, I realized that the mind is not confined to the physical brain. How could it be when the ether around synesthetes dances with these colorful motile things in response to sensations? Sometimes synesthesia feels like a dimension apart.

Among synesthetes, there are "associators" and "projectors." Associators experience the association as a thought or see it in the mind's eye. Projectors see things in near space around their bodies. Some people experience both.

Where is this synesthesia space that overlaps consensus reality? And what can it tell us about all our sensory systems and their physics?

Dr. Eagleman told me about a South Asian research subject who sees the names of the categories of India's caste system all around him—but not in the normally imposed order. Brahmins are not higher than Untouchables in his experience, and the castes do not appear as a list; the words are splayed out everywhere.

I asked the synesthetes who meet in various corners of the internet to share their sense-of-place experiences. Jade McLeod had some of the most provocative. I found her in one of my favorite gathering spots, the Facebook group I Have Synesthesia. I'm Not a Freak, I'm a Synesthete.

McLeod, an artist and special education teacher, said she has synesthesia in response to orgasm. This does not happen every time, she said, only when she is very relaxed. I asked her what the related imagery looks like.

"Nothing in the room is actually still visible," she told me of the mind-space.[5] "It's like a dream-state sort of takes over." At first, she said, it's just "flashes of things, almost like a movie in my mind. And then it's just like a complete takeover. It goes from rapid visions to a complete transformation." She said the feeling is one of full immersion. "I mean, I'm aware of the source. But everything else disappears. It could be a desert-scape, lightning, et cetera. But the colors are almost always violet, blue, kind of electric-looking."

Bradford Keeney, a self-described "recovering psychotherapist" who is recognized as a shaman in many cultures around the world, taught me about the Kalahari Bushmen and their synesthesia "place."

The Kalahari Bushmen have the oldest surviving culture in the world, with sixty thousand years of history, and they painted synesthesia on cave walls, he said. Keeney sent me photos of the art, and the figures of people and animals appeared to be surrounded by synesthesia photisms; neuro-typical people might see them as clouds.

Keeney's time with the Kalahari Bushmen shamans led him to believe every person can access synesthetic experiences. He told me in a previous interview it is also evident in the language of spiritual elders he has encountered throughout the world. "Strong spiritual elders talk this way. They can smell another spiritually developed person, while seeing their light and hearing their song.... You are experiencing everything at once with no need to distinguish. The highest spiritual experiences feel, see, hear, taste, and smell at the same time without conscious differentiation. It is synesthesia with no conscious narration about it being synesthesia. No distinction. Only whole experience. We learn to draw distinctions and make indications that then enable us to say, yes, I am smelling love and hearing hope."[6]

Many creative people have spoken with me over the years about their synesthesia experiences. Virtuoso actor and synesthete Geoffrey Rush has won almost every major acting award from Emmys to Oscars. But the synesthesia impressions that surround him when he performs are even more beautiful than any of the theatrical sets he has graced. He told me that when the Western calendar year changed from 1999 to 2000, he experienced a change in color. "I suppose because [in 1999] there were three nines and it was saturated, and nine for me has always got a sort of deep violet-lavender aura to it—and then 2000 was almost completely pure white because of all those zeros, I suppose. And the colors are never like paint-chart colors."[7]

Legendary violinist Itzhak Perlman has an inner life as mesmerizing as the sweet tones of his 1714 Soil Stradivarius. He describes certain sounds with color. "It's not music—it's notes, it's single sounds." For instance, if he plays a B-flat on the G string of his violin, the color is a deep forest green. "And if I play an A on the E string, that would be red. If I play the next B, if I look at it right now, I would say that it's yellow. The bright colors are the upper strings of the violin—for me. I associate it with bright colors of the spectrum."[8]

I met Perlman backstage at Lincoln Center after his performance one evening, and when I handed him a bouquet of red roses, he joked that the red roses were all wrong because *rose* is a blue word to him. I told him it was a terracotta word to me.

Billy Joel, who is integral to the Great American Songbook, like Perlman, experiences waves of color in the space around him when he composes or performs. For my first book, *Tasting the Universe,* he told me, "When I think of different types of melodies which are slower or softer, I think in terms of blues or greens." Songs that had a heavier beat and a faster rhythm suggested the red-orange-yellow end of the spectrum for Joel (think "It's Still Rock 'n' Roll to Me" and "We Didn't Start the Fire").[9]

There are more than one hundred varieties of synesthesia cataloged to date, and while these artists and I all experience colorful impressions in response to music, none of the experiences are exactly the same. For Perlman, it is a note-to-color association; for Joel, it is genre-to-color; and for me, it is timbre-to-color. (In other words, the colored forms I see depend on which instrument is playing.) And the numbers of the new year went from mostly onyx to mostly white for me—as my nines are black and my zeros are white.

More and more creative synesthetes are going public with their traits in this new climate—Mary J. Blige, Grimes, Dave Grohl of the Foo Fighters, Alanis Morissette, LL Cool J, and Lady Gaga are just a few. Appearing on *The Tonight Show Starring Jimmy Fallon,* synesthete Billie Eilish explained her signal-before-the-screen experience. "For instance, every day of the week has a color, a number, a shape. Sometimes things have a smell that I can think of, or a temperature, or a texture." Eilish's synesthesia is chronicled in her work; if you go to her concerts, you'll see her "synesthesia place."

"All of my videos for the most part have to do with synesthesia. All my artwork, everything I do live, all the colors for each song, it's because those are the colors for those songs," she told Fallon.[10]

Eilish created an immersive synesthesia exhibit in Los Angeles to launch her first album, *When We All Fall Asleep, Where Do We Go?*

"I wanted them [her fans] to experience it in all aspects," she told *Billboard*'s Chelsea Briggs at the opening, "instead of just hearing my album and you're done, and you go home. I wanted it to literally be like an exhibit, a museum, a place to smell and hear and feel. Every room has a certain temperature, every room has a certain smell, a certain color, a certain texture, a certain shape, a certain number on the walls. And it's all like synesthesia in my head and that's how I create music in the first place."[11]

Synesthetes have also been reaching back in time to find evidence of historical synesthetes.

Norman Mailer wrote about Marilyn Monroe's bonus senses in *Marilyn: A Biography*, in 1973, although he didn't use the word *synesthesia*. He recounted the recollections of Monroe's first husband, Jim Dougherty, who remembered that there were "evenings when all Norma Jean served were peas and carrots. She liked the colors. She has that displacement of the senses which others take drugs to find."[12]

I found Mona Rae Miracle, Monroe's surviving niece and biographer, and asked her about Monroe's possible synesthesia. She said not only had Monroe experienced it, but she did too. Synesthesia has a genetic basis, and it tends to run in families. When I wrote about Monroe's synesthesia in my *Psychology Today* blog, the singer Lorde, also a synesthete, tweeted about it.

Puerto Rican blogger and synesthete René D. Quiñones reached back in time and found that Vincent van Gogh was a synesthete. In 1881, when Van Gogh was twenty-eight and living in The Hague, he wrote his brother and benefactor, Theo, about his work with lithographs. Quiñones dug up the letter from a research archive.

"Some time ago you rightly said that every colorist has his own characteristic scale of colors," Van Gogh wrote. "This is also the case with Black and White, it is the same after all—one must be able to go from the highest light to the deepest shadow, and this with only a few simple ingredients. Some artists have a nervous hand at drawing, which gives their technique something of the sound peculiar to a violin, for instance, Lemud, Daumier, Lançon—others, for example,

Gavarni and Bodmer, remind one more of piano playing. Do you feel this too? Millet is perhaps a stately organ."[13]

This form of the trait is very rare—a type of *technique-to-timbre synesthesia.*

Van Gogh's experience makes me wonder if he was overwhelmed by such impressions. They can be very over-stimulating at times, and there is no way to turn them off.

Antiques dealer Lynn Goode of Houston got a glimpse of what it is like to live without multiple forms of synesthesia when she came down with COVID. Synesthetes sometimes lose their impressions in times of illness or stress. "There was one positive side effect from losing synesthesia for those few months: there was not the usual tsunami of senses that occur in my Technicolor brain. On the few times I ventured out of the house to pick up necessities, I wasn't bombarded with what I now understand as a neurosensory overload whenever I entered an unfamiliar place. With synesthesia, my ability to process neural sensory stimulation is involuntary, which means my neurons seemingly take in everything at once. The best way I can describe this nervous sensory overload is it's like being the subject of a flash photo: I'm momentarily blinded by the flash, and it takes a moment to process all that I am seeing/feeling/hearing in those first seconds."[14]

When her synesthesia came back, her neural pathways seemed to have reset. "Like after a rainstorm clears, my senses seem much brighter now that they've returned. I wonder what happened neurologically to cause the blackout and subsequent relighting of my senses, and further if it can be stimulated in non-synesthetes? Or conversely, if there is a way to turn off the overfiring of various synapses?"

We modern synesthetes can speak openly about the experience without being accused of hallucinating, tripping on LSD, or seeking attention, and this is partly thanks to brain-scan studies conducted in the 1980s by Dr. Cytowic at George Washington University and fellow father of modern synesthesia research Dr. Larry Marks of Yale University. Though their peers told them the work was too woo-woo and could end their careers, these heroes of neurodiversity persevered and proved synesthesia was real.

Synesthesia wasn't always reviled or ridiculed. A hundred years before Cytowic and Marks came along, the trait was well known and considered chic; it was represented in performances of lighted color organs, which had a different color for every note.

Synesthesia has once again become a desirable trait. *Bloomberg BusinessWeek* noted this a couple of years ago in a piece titled "The Mind's Eye: Synesthesia Has Business Benefits." The article—which unfortunately, and mistakenly, referred to synesthesia as a "mental condition" or "hallucinations"—featured prominent synesthetes like Michael Haverkamp, an engineer for Ford Motor, and UN instructor Patricia Lynne Duffy. Haverkamp said he optimized car design by using the synesthetic reactions he got from touching automobile features. Duffy pointed out that it was easy for her to remember people's names (the ultimate act of diplomacy) because of the added synesthetic colors she sees for the letters. And synesthete and olfaction expert Samantha Goldworm shared the benefits of her trait in creating olfactive landscapes for major designers.[15]

In my previous book, *Struck by Genius*, I profiled the extraordinary acquired savant and synesthete Jason Padgett, who can see the fabric of things.

Synesthesia gave me eyes to see Jason. I discovered him through a Google alert that I had set up for anyone posting about bonus senses. The YouTube video he'd uploaded of his drawings of the geometric forms he sees synesthetically had only forty-three views at the time, but I knew within seconds that his visions were something very rare indeed. My own synesthetic forms (photisms) and those I've seen represented in other synesthetes' artwork are unforgettable to me, so I was immediately interested in the variations and greater complexity of his synesthetic place.

It also gave me eyes to see the synesthesia rendered in the work of Swedish painter Hilma af Klint when I went to an exhibit of her work at the Guggenheim Museum. Her place looked a little like my place. I later confirmed with one of her surviving relatives that she'd experienced synesthesia.[16]

Many countries now have organizations devoted to the bonus sensory set. The latest is the Synesthesia Society of Africa. Founded by Abiola Ogunsanwo of Nigeria, the group is a victory for pan-African neurological outliers who have until now been underrepresented in research, the arts, and related synesthesia business.

Board member Dr. Sheila Clare Butungi is a veterinarian in Uganda. She sought out the online community, as she did not know any other synesthetes in Entebbe.

In many ways, synesthetes feel more connected to other synesthetes than they do to people who share their ethnicities or nationalities. I've seen friendships form in this neuro tribe and cross borders many times.

The late U.S. Army lieutenant colonel and synesthete Jim Channon was the commander of what was known as the "Men Who Stare at Goats" unit at Fort Bragg. His research was immortalized by author Jon Ronson in a bestselling book. Lieutenant Colonel Channon had been greatly influenced by the human potential movement in California in the 1970s before applying it to the military. His mastery of esoteric knowledge and his business sense earned him the title "the first corporate shaman" by *Fortune* magazine.

I talked with him about synesthesia. He called it "poly-experiential awareness" and likened it to a "twelve-D world," also giving it a sense of place. After that conversation, he shared this:

"Tend to your dreams with more reverence. Start to add qualities to *all* your feelings. Ask your nose for more detailed information. Distinguish when your memory and that subtle voice inside are two different sources. Make the touch you give to others a language you can modify and perfect. But *mostly* stop believing that the only ways that are real are the ones we can speak about."[17]

If it sounds far out, he *was* far out. And so is synesthesia. Synesthesia feels like it is happening on the same minute scale science is proving in the individual senses—and in another dimension. Eleven years ago, in *Tasting the Universe*, I theorized that synesthesia happens on a quantum level because of its interdimensional quality and that it is a special form of consciousness. One of the people I interviewed for that book, noted writer, artist, and synesthete Doug Coupland, had this to say:

"Funny you should mention quantum consciousness. I kind of arrived at the possibility of its existence last year while reading Jeff Hawkins's *On Intelligence*.... I've been concluding that life is simply nature's way of crossing distance, and that life is a sort of Rube Goldberg affair that atomic structure leaves no choice but to create the first chance it gets. What we call *perception* is merely the localized form of this within our own bodies."[18] Nobel Prize winner Sir Roger Penrose is a big advocate of consciousness research. He doesn't experience synesthesia himself, "but I have found it to be fascinating," he told me. "Why should it make a difference to one's actual perception which particular neurons are being activated?"[19]

Synesthete artist and graphics professor Carrie C. Firman articulated it well: "Most neurodivergent people become so used to masking and acting like we are not picking up on the quantum that we can lose track of where we are and where our masks are. Often, we withdraw due to misunderstanding and the burnout that trying to be understood causes. We literally operate at a different level."

Western science is obsessed with the observable, Firman said. "'I'll believe it when I see it.' Eye roll." She thinks that people with more highly attuned senses can pick up and interpret things between quantum and wherever the heck the bar is for "observable" better than "the regular people" can.

She dated a man with ADHD for about a year. "I learned so much. He was much more verbal about expressing his experience and frustrations, probably because he was diagnosed as a child (very easy for a white male). One of his biggest frustrations was honoring his brain and having people use phrases like 'too sensitive' with the intent of emasculating him. It's not surprising that we had amazing chemistry, similar experiences, and very special points of connection." They both understood the power of being able to sense what most others in society couldn't.

Firman said TikTok has been a great resource; when neurodivergent people tell their stories and relate it back to research, it makes her feel seen. "One of the things that will pop up is the efficiency and intensity of two neurodivergent people who occupy similar wavelengths . . . the fact that our home territory is under the radar enhances the connection."

Our home territory.

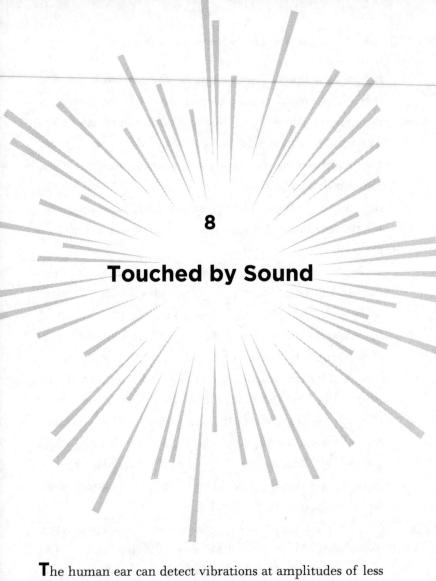

8

Touched by Sound

The human ear can detect vibrations at amplitudes of less than the diameter of an atom and in intervals as short as ten-millionths of a second. Until recently, knowledge of the auditory cortex was based primarily on animal studies, but in 2008, during a clinical procedure to evaluate neural pathways in patients with epilepsy, scientists were able to record the activity of single neurons in the human auditory cortex for the first time.[1] They played random chords and found that subjects could discriminate between sounds better than most mammals, detecting frequency differences as small as a quarter of a tone. Interestingly, when the subjects heard real-world sounds—such as dialogue and music from the soundtrack of *The Good, the Bad, and the Ugly*—the auditory cortex processed the information differently, suggesting that the brain tunes in to different frequencies based on the context of the situation.

In 2018, speaking at the second International Conference on Vajrayāna Buddhism in Thimphu, Bhutan, William C. Bushell said, "We again encounter the comparison of the human sense organs to highly sophisticated, advanced, state-of-the-art human-made technological apparatus."[2] Two of the finest minds in the field, A. J. Hudspeth of the Rockefeller University and Tobias Reichenbach of Imperial College London, noted, "The performance of the human ear would be as remarkable for a carefully engineered device as it is for a product of evolution. . . . Explaining how the ear meets these technical specifications is a major challenge for biophysics."[3]

Two other leaders in the field of auditory psychophysics, Meredith LeMasurier and Peter G. Gillespie, agree: "The performance of the mammalian auditory system is awe-inspiring. The cochlea, the organ responsible for auditory signal transduction, responds to sound-induced vibrations and converts these mechanical signals into electrical impulses, a process known as mechanoelectrical transduction. . . . The hallmarks of cochlear transduction are incredible sensitivity, versatility, and speed."[4]

It really is astonishing. The human ear can respond to the largest range of stimuli of any of the senses; it can detect frequencies from as low as 20 to as high as 20,000 hertz.

Twenty hertz would be what you'd hear "if you were at an R&B concert and you just stood next to the bass," said Michael Qin, a senior research scientist at the Naval Submarine Medical Research Laboratory in Connecticut, speaking to *National Geographic*.[5] "It would be the thing that's moving your pants leg." On the high end of the range, 20,000 hertz—or 20 kilohertz—would sound like a very high-pitched mosquito buzz.

Humans can detect signals with intensities that are less than one-billionth that of atmospheric pressure. Perhaps that is why percussionist Dame Evelyn Glennie, who is deaf, says, "Hearing is a form of touch. You feel it through your body, and sometimes it almost hits your face."

One of the most amazing examples of human hearing I've ever encountered is Brian Wilson of the Beach Boys. He has been deaf in his right ear since childhood. Yet consider the soaring harmonies he helped create for the band. "Jesus, that ear. [Wilson] should donate it to the Smithsonian," Bob Dylan said about Wilson's intact and miraculous left ear.

I interviewed Wilson in 2017 prior to a concert at Manhattan's Beacon Theatre. I wanted to know how he managed to be such a "golden ear"—someone with super-hearing— when he had only one functioning organ.

"A professional musician who can't hear on one side of his head? It's funny but not funny," he wrote in his memoir *I Am Brian Wilson*. "Over the years I have learned how to make it work in the studio but it's harder onstage, where you have to know what's going on all around you. It's hard to stay on key when you can't hear everything that is being played. The sound up there can be overwhelming, and I have only one monitor, to my left. It has to be positioned perfectly, just right, or all I can hear is noise."

I mentioned to Wilson that deaf percussionist Dame Evelyn Glennie once told me she takes her shoes off to "hear" her accompanying orchestra through the stage with her feet.

"Yes, I do hear the music that way! Through my entire body!" he exclaimed.[6]

The sensoria do not act separately but as a whole, compensating for deficits in some channels by using others. In their 2014 paper "Compensatory Plasticity in the Deaf Brain: Effects on Perception of Music," researchers Arla Good, Maureen J. Reed, and Frank A. Russo of Ryerson University in Toronto wrote: "When one sense is unavailable, sensory responsibilities shift, and processing of the remaining modalities becomes enhanced. This shifting of responsibilities appears to be compensatory in nature and has thus come to be referred to as compensatory plasticity."[7]

When you learn how the ear works, you realize, once again, that humans are very "soft-tissue/high-tech." Hearing is a multistep process. The National Institute on Deafness and Other Communication Disorders explains it this way:

- Sound waves enter the outer ear and travel through a narrow passageway, the ear canal, to the eardrum.

- The eardrum vibrates in response to the incoming sound waves and sends these vibrations to three tiny bones—the malleus, incus, and stapes—in the middle ear.
- The middle ear amplifies, or increases, the sound vibrations and sends them to the inner ear, which houses the cochlea, a snail-shaped structure. An elastic partition, called the *basilar membrane*, runs from the beginning to the end of the cochlea, splitting it into two fluid-filled chambers.
- Vibrations cause the fluid inside the cochlea to ripple, and a traveling wave forms along the basilar membrane. Hair cells—sensory cells sitting on top of the basilar membrane—ride the wave. Hair cells near the wide end of the snail-shaped cochlea detect higher-pitched sounds, such as an infant crying. Those closer to the center detect lower-pitched sounds, such as a large dog barking.
- As the hair cells move up and down, microscopic hairlike projections (known as *stereocilia*) perched on top of the hair cells bump against an overlying structure and bend, creating an electrical signal.
- The auditory nerve carries this electrical signal to the brain, which turns it into a sound that we recognize and understand.

This amazing sound system got its start in the bony gills of fish, according to research conducted in 2006 at Uppsala University in Sweden.

In the journal *Nature*,[8] scientists Martin D. Brazeau and

Per E. Ahlberg presented evidence indicating that the bones of the human ear evolved from the breathing tubes of ancient fish. The researchers examined a 370-million-year-old fossil of a fish called *Panderichthys* and found that its spiracle tract (the structure ancient fish used to breathe) strongly resembled the stapes-like bone found in the middle ear of early vertebrates, suggesting that, evolutionarily, the middle ear was once part of the breathing apparatus, although it's unclear how the structures went from helping creatures breathe to helping creatures hear.

We need to take care of these complex sound systems, nearly four hundred million years in the making, just as we do our high-end electronics, Grammy-nominated composer Steve Roach, who is a synesthete and a golden ear, said. Roach orchestrates stadium shows and other performances as well as the sound in his home environment. He said it's very important to create a beautiful soundscape where you live and work. You can cancel a lot of annoying and sometimes dangerous ambient noise with a good sound system and playlist, he said, speaking to me from his ranch and home studio in Tucson, Arizona. There, he and his wife, Linda Kohanov, create a healthy ambient-sound environment.

"Once you find that sweet spot, you can loop these days with all the sources of digital music at any moment through your phone or iPad or computer. You can find that music and build a playlist. I have a playlist on Spotify that's called 'Mythic Imagination,' and it plays for ten or twelve hours."

He believes sound can be nurturing and restorative. "I like to call it *recalibrating*, because you're bringing it down to where you're not demanding all this attention on your ears, because so much in our world now is demanding the attention of our senses. The visual is just like overload. The sound is overload."

Roach says hearing nurturing sounds is as essential as eating the right food. "You're going to choose what you eat every day. You should also choose what you listen to." When you don't pay attention to your auditory environment, you can get into trouble. It's almost like the frog in a pot of water that is slowly brought to a boil—the frog doesn't realize the danger until it's too late. "The [sound] level just snuck up on us, where you just don't realize at a certain point how loud the world has become, how dense the sound field has become," Roach said. "It's amazing how people can live in it—there are these eight-lane highways with condos right next to them."[9] For more on these issues, he recommends the book *Noise* by Bart Kosko.

We've seen how extraordinarily sensitive and powerful hearing is, with the widest range of any of our senses. And according to an article published in the journal *BMC Public Health*, stress can cause hearing loss: "The study unambiguously demonstrates associations between hearing problems and various stressors."[10]

Meditation may reduce the stress response and decrease the loss of hearing; it may even improve it. According to several studies, meditators have increased tissue in parts of the brain responsible for attention. *The Hearing Review* reported that "an increase in these areas would suggest that you would be better at paying attention if you meditate, but it's not just focusing. Meditating increases an individual's ability to perceive sounds at different auditory thresholds and increases the way the brain codes and stores this auditory information."[11]

We would all like to have healthy hearing, but as Dr. Bill Budd of the University of Newcastle in England told ABC News, you don't necessarily want the gift of super-hearing. "Hearing isn't like other senses; if a light is too bright, we can always close our eyes or turn away, but our hearing is always 'on,' even when we are asleep, and super-sensitivity would be quite impairing," said Budd.[12] Wearing earplugs is not always practical, since there are sounds you need to hear to alert you to danger. If we all had super-hearing, "we would be surrounded by such a cacophony of sound we wouldn't hear anything very well."

People with golden ears are usually very sensitive to a particular aspect of sound. Some may hear very soft sounds well; others may be good at hearing high-frequency sounds; still others may be good at determining tiny timing differences; but very few golden ears have better-than-average abilities in all those ways.

"This is actually like thinking about high-level athletes," sensory researcher Kristopher Jake Patten of Arizona State University told me.[13] "Us mere mortals can never run as fast as, say, Usain Bolt. It may be true that Bolt, like tetrachromats, has something fundamentally different about his physiology than the rest of us. But that doesn't mean he doesn't train. A talented artist may have better muscular control over their fingers, but they also draw frequently. Musicians are a particularly good case study for this. There are people with perfect pitch who can detect when a note is mistuned by a minute amount. Without musical training, a person with perfect pitch would always outperform a person with typical pitch perception in a test of identifying when two pitches were matched or mismatched. With musical training, however, musicians with perfect pitch do not perform significantly better than their non–perfect pitch counterparts. Training, then, is much more important."

To maximize sensory potentials, Patten advised, use the sense and direct your attention to it. "If you want to be better at identifying pitch differences, listen to more and varied kinds of music. Or, better yet, start playing an instrument. To better discern the differences between colors, play with color by painting or designing something. Try different food and try to taste all the individual ingredients or focus on the feel of different textures as you run your hand over them. Our sensory apparati and brains are amazing and, when we train ourselves to pay attention to and appreciate differences, we can unlock richer worlds."

The extreme sensitivity of hearing helps us to discard myths about our abilities. To move into a more discerning sensory future, we need to discard many more myths, such as the myth that there are only five senses, a theory put forth by Aristotle in *De Anima.*

I bring this up here because the ears are also essential to another sense we have—balance, or equilibrioception. The inner ear helps tell the body where it is in space at any given time and adjust for balance. Retired U.S. Air Force major Randy Eady, a physiologist and balance expert, explained that equilibrioception requires several senses working at once. The vestibular, ocular, and proprioceptive systems are very important in everyday life. Living without one of the three is doable, though challenging. But, Eady said, "If only one of the three systems is functioning, kiss your balance goodbye." That's because "a once-automatic reflex, balance, now may require great concentration. The area of the brain once reserved for memory and thought processing must now focus on balance control."[14]

Eady, in cooperation with numerous collegiate and professional ball-sport teams, developed and applied a program at the Air Force Academy that used natural eye movements and focus techniques to aid in balance. The approach works by creating pattern recognition of movement.

Neuroscientists put the number of senses anywhere from twenty-two to thirty-three, said Barry C. Smith, director of the Institute of Philosophy at the University of London's School of Advanced Study. He is also the founding director of the Centre for the Study of the Senses at the school, which pioneers collaborative research between philosophers, psychologists, and neuroscientists.

"We still are in the grip of an Aristotelian view of our senses," Smith said in a video for *Aeon*.[15] "If we think there are five senses, and we think that they are working independently of one another, as we've done for about two and a half thousand years, we may be starting with the wrong premises. . . . They don't come in separate packets and there's no neat line distinguishing them."

Some of the other senses, according to the Sensory Trust in the UK, are:

- Thermoception—the sense of heat (some researchers believe that the sense of cold may be a separate sense)
- Nociception—the perception of pain
- Equilibrioception—the perception of balance
- Proprioception—the perception of body awareness (such as being able to touch your nose with your fingertips with your eyes closed)[16]

"Our senses work together, and what one sense perceives can manipulate what another perceives," Smith continued. He gives an example of an airplane: when you're sitting in a plane, what you see changes as the plane takes off—the cabin in front of you looks higher, even though nothing has actually changed in our field of vision. That's because your inner ear is telling you that you're tilting backward, so that changes what you see.

Our sense of hearing is perhaps the most extraordinary sense of all. To celebrate this, I've made a Spotify playlist with the same title as this book. I hope it provides an eclectic ocean of sound for you to enjoy as you read.

The myth that there are only five senses and that they operate separately persists, Smith said in *Aeon*, "perhaps because a clearer understanding of our sensory experience at the neurological level has only recently started to take shape."

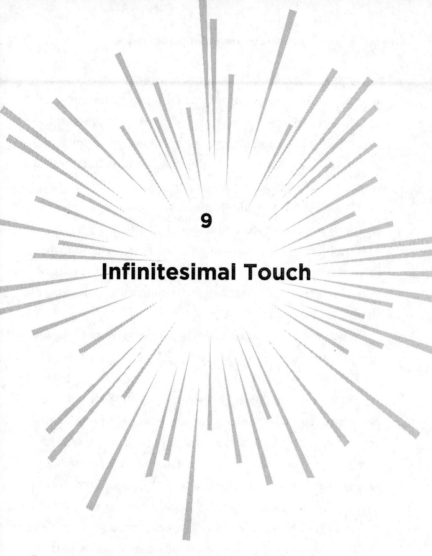

9

Infinitesimal Touch

"I have this pet peeve about smartphones," David Howes said, swiping his hand across the screen of his own mobile device to demonstrate.

I introduced you to Professor Howes, the anthropologist, author, and pioneer of sensory research, in the olfactory and mirror-touch synesthesia chapters. He directs the Centre for Sensory Studies at Concordia University in Montreal, an extraordinary realm of next-level thinking on the topic.

The smartphone "is so devoid of sensations," he continued. "One of the most satisfying experiences is one of those old rotary dial phones where the number nine has a lot of resistance, and it takes time for it to come back."

I thought of the old black desktop Western Electric phone in my childhood home and of dialing my grandparents' number, which had a nine in it. It was so delightful, in retrospect—the tug of the dial, the shiny enamel, the sound. It gave you time to think before you spoke and a chance to clear your throat.

When the world is too smooth, people do bumpy things like get tattoos, Howes said.

Wait. What?

As an anthropologist, Howes takes a holistic view of behaviors and how they are interrelated. He looks at the spread of tattooing and thinks, *Why?* "Tattooing was formerly a very antisocial practice. The only people who got tattoos were prisoners and sailors, both people on the outside of society," Howes said. "Now it has become mainstream. Why? Because we're sensation-starved."[1]

The physical feeling of being tattooed is a large part of the appeal, Howes said. "And it also enables you to individually mark yourself off from others by your skin.

"We live in a culture of smooth sensations. Anesthesia is very important—feel no pain." And yet there are "grinders," Howes pointed out. He was referring to another example of pushback against a world grown too smooth, biohackers "who use magnetic implants to achieve a more-than-human

sensorium." In a *New York Times* article about the subcul-
ture that's working to expand the senses, one grinder described an implanted finger magnet as creating a feeling
"like the air has a texture."[2]

Lack of touch is a huge social and health problem. One
of the most popular features on *Psychology Today*'s website
in 2013 was a piece on "skin hunger" and how many people
crave the touch of another human being. It made me think
of the Romanian children who lived in orphanages where
they were hardly ever touched and who suffered significant
physical and mental developmental delays because of it.[3]
"Touch comes before sight, before speech," Margaret Atwood
said. "It is the first language and the last, and it always tells
the truth."[4]

"Just as lack of food, water, and rest have their detri-
mental effects," wrote Kory Floyd, professor of health and
family communication at Arizona State University, in *Psychology Today*, "so too does the lack of affection. In a recent study of 509 adults, I examined the construct of skin
hunger—and the social, relational, and health deficits with
which it is associated." Floyd observed, "The results were
consistent and striking—the study found that people with
high levels of skin hunger were disadvantaged in multiple
ways when compared to people with moderate or low levels.

"Such people are less happy, more lonely, more likely to
experience depression and stress and in general, in worse
health. They experience more mood and anxiety disorders,
and more secondary immune disorders." They were also
"less likely to form secure attachments with others in their
lives."

His findings don't establish that skin hunger causes all these negative conditions, he pointed out, "only that people who feel highly affection-deprived are more likely than others to experience them."

He concluded: "Fortunately, skin hunger doesn't have to be a permanent condition. Each of us has the capacity to get more affection in our lives. In the meantime, *put down your cell phone* and share an affectionate moment with someone in person. For those with skin hunger, human contact—not the technologically mediated variety—is the cure for what ails."

Dacher Keltner, a professor at the University of California at Berkeley, and his student Matt Hertenstein measured sensitivity to touch in the laboratory and found that subjects could even discern the emotion of the person touching them.

"We built a barrier in our lab that separated two strangers from each other," Keltner wrote in an article in *Greater Good* magazine.[5] "One person stuck his or her arm through the barrier and waited. The other person was given a list of emotions, and he or she had to try to convey each emotion through a one-second touch to the stranger's forearm. The person whose arm was being touched had to guess the emotion."

A number of emotions were considered, and the odds of guessing the right emotion purely by chance were about 8 percent. Remarkably, though, when the emotion being conveyed was compassion, participants guessed correctly nearly 60 percent of the time. And they got gratitude, anger, love, and fear right more than 50 percent of the time.

This means perception through our sense of touch is stronger than visual cues—like the emotion on another person's face—and even verbal communication.

"Regrettably, though," Keltner said, "some Western cultures are pretty touch-deprived, and this is especially true of the United States." In many other countries, "people spend a lot of time in direct physical contact with one another—much more than we do."

One of his favorite examples of this is a study done in the 1960s by pioneering psychologist Sidney Jourard. He and his team observed the conversations of friends in different parts of the world while they sat in a café together.

Jourard found that in England, the two friends touched each other zero times in an hour. In the continental United States, people touched each other twice. But in France, the number shot up to 110 times an hour, and in Puerto Rico, friends touched each other 180 times an hour.

Your skin, the largest organ of the human body, knows what active and passive touch is. You are, in fact, touching something all the time—the chair you're sitting on, the floor beneath your feet—but it does not substitute for the touch of another human.

"You can't turn off touch. It never goes away," David Linden, a neurobiologist at Johns Hopkins and author of the book *Touch: The Science of Hand, Heart, and Mind,* told *Vox* recently.[6] "You can close your eyes and imagine what it's like to be blind, and you can stop up your ears and imagine what it's like to be deaf. But touch is so central and everpresent in our lives that we can't imagine losing it."

There are different types of receptors in the skin.

Thermoreceptors send signals about temperature, *nociceptors* send signals about pain, and *mechanoreceptors* send signals about physical changes. Within each category, different types of receptors detect different stimuli. For instance, one mechanoreceptor senses vibration, one senses the stretching of the skin, and one senses very light touch. Receptors of this last type are called *Merkel endings,* and they're found, according to David Linden, "only in the parts of your body you use to feel something really finely—like your fingertips and lips."

And there are two touch systems, Linden said. One "gives the 'facts'—the location, movement, and strength of a touch." The other system is for emotional touch. "It's mediated by special sensors called *C tactile fibers,* and it conveys information much more slowly," he said. "It's vague—in terms of where the touch is happening—but it sends information to a part of the brain called the *posterior insula* that is crucial for socially-bonding touch. This includes things like a hug from a friend, to the touch you got as a child from your mother, to sexual touch."

Touch has multiple moving parts, and it's orders of magnitude more sensitive than most people realize. In 2017, a team of University of California at San Diego researchers led by Darren Lipomi tested whether human subjects could, using only their fingers, distinguish between smooth pieces of silicon that differed only in their single topmost layer of molecules.[7] One surface was a single oxidized layer made mostly of oxygen atoms, and the other was a single Teflon-like layer made of fluorine and carbon atoms. Both silicon surfaces looked the same, and they felt similar enough that some participants in the study couldn't tell the difference. However, others were indeed able to discriminate between these two surfaces that differed by just a single layer of molecules.

"It was definitely surprising to us how sensitive human fingertips are to such minute differences," Lipomi told me. He agreed that all the new sensory research, including that at his own laboratory, might amount to a movement.

"We might be seeing a Renaissance. I think the most interesting science happens in the space between disciplines separated by a lot of intellectual distance. I am an organic chemist and materials scientist, but my most interesting collaborations right now are with behavior and cognitive scientists."

Lipomi has already proven things at a molecular scale, so could quantum touch be next? "We don't have plans to look at the quantum level, but in a way, any interaction of a fingertip with a surface is already influenced by van der Waals force, which has its origins in quantum mechanical fluctuations of electron density," he told me.[8]

Up to now, the sense of touch has been the most neglected sense, but inspired by this newly discovered human potential, researchers are now working on everything from tactile therapy for newborns to haptic suits for virtual-reality gaming to surgical robots and wearables for people with neurological disorders.

"We are interested in how the tools of materials science can be used to understand the human sense of touch," Lipomi said. "Ultimately, we hope to use these techniques to develop haptic interfaces for education and training, enhanced surgical procedures, and tactile therapy."

Some people on the autism spectrum have a hypersensitivity to touch. While this is an extraordinary super-sense, it makes life very difficult for them and those who care for them. Some dislike the feeling of clothing against their skin; some refuse to be held or touched and hate having their hair attended to. In severe cases, children with autism even refuse to swallow food.

A study conducted at Harvard University suggested that autism may not be a brain issue so much as a touch issue. Researcher Lauren L. Orefice reported in the journal *Science* in 2019:[9]

"Autism spectrum disorders (ASDs) are thought to arise exclusively from aberrant brain function. Our research proposes a surprising revision of this view. We have discovered that peripheral sensory neurons—neurons outside the brain—are key sites at which ASD-related gene mutations have a critical impact." Researchers found that, in mice, dysfunction of certain peripheral neurons disrupted central nervous system development and caused ASD-like characteristics, such as sensory overreactivity, social impairments, and anxious behaviors.

David Moore, a senior lecturer in psychology at Liverpool John Moores University in the UK, believes that the scientific community is too focused on the brain and insufficiently focused on the parts of the body that influence the brain.

I asked Moore to elaborate. "The brain is clearly important in processing and interpreting the information that we experience; however, it is our periphery that is out there accessing these inputs," he told me. "Our eyes allow us to see, and without them, our brain could not interpret vision. I am most interested in how we experience touch and pain. The threshold at which our peripheral fibers are activated matters for what we feel and when we experience pain." When researchers measure the brain only, they're considering the outcome of the experiences but ignoring the genesis of them.[10]

Haptics—the science of touch—is a very hot area in sensory research right now. Scientists are focusing on interfaces that will help people experience touch from remote or simulated environments. However, given the astounding sensitivity and complexity of human touch, anything realistic or meaningful in this department is most likely a long way off.

Sensory impressions don't always rise to the level of someone's conscious awareness until someone else points them out. Perhaps that's what makes them easier to ignore and dismiss on a societal and scientific level. Taking a conscious inventory of your sensory input and then doing the work to utilize the senses in a more deliberate way can give you a more powerful experience of the world.

The senses are a bottom-up group of channels. Perception is top-down—that is, the mind understands and utilizes impressions to create the collage that forms reality. Or perhaps it is not so much a collage as an interwoven tapestry.

Neuroscientist Charles S. Sherrington wrote about this forming of reality in his 1942 book *Man on His Nature*. He famously called the brain "an enchanted loom":

"The brain is waking and with it the mind is returning. It is as if the Milky Way entered upon some cosmic dance. Swiftly the head mass becomes an enchanted loom where millions of flashing shuttles weave a dissolving pattern, always a meaningful pattern though never an abiding one; a shifting harmony of subpatterns."

There is a concept in neuroscience and psychophysics called "the absolute threshold of sensation."[11] That threshold is the level at which a stimulus—be it the smell of cinnamon, the sound of thunder in the distance, or the feel of a breeze—will be detected by an organism at least 50 percent of the time. The absolute threshold of sensation used to be defined as the lowest level of a stimulus that an organism could detect, but it was redefined because other factors—the subject's environment, expectations, and experiences, for instance—alter how much attention is paid to the stimulus. (Parents attending to a colicky baby may not hear the doorbell ring and not just because the baby is loud but because the baby's care is their priority.) And there's the crux of the matter. It's about attention and focus and opening oneself up to one's potential.

People are seldom still and attentive enough to give much thought to the senses beyond the most pressing—swatting a mosquito that has landed on one's arm, for instance. We're cut off; we're "out of touch," as the saying goes.

If we were taught how to use the senses, we might enjoy life on a deeper level. We know now there is much to be gained. Our reality is formed by the senses. Perception can be a barren landscape or a lush, vital, and informative one. Where would we rather live?

Psychological studies have found that people who expect to succeed at whatever task they're working on do tend to succeed more often. I asked the prominent Harvard University psychologist, author, and intellectual Steven Pinker about this. If people knew they had such extraordinary sensory potentials, would they be more likely to develop them? "Fascinating, Maureen," he replied in an email. "Yes, I think it could inspire an appreciation of the amazing abilities that evolution installed in us."[12]

He directed me to the work of Carol Dweck at Stanford University. Dr. Dweck is a psychologist who studies why and how people succeed. Her conclusion, after decades of research into achievement, is that "in a fixed mindset, people believe their basic qualities, like their intelligence or talent, are simply fixed traits. They spend their time documenting their intelligence or talent instead of developing them. They also believe that talent alone creates success— without effort. They're wrong," she wrote in *Mindset.*[13]

People have a fixed mindset about the senses, I believe.

She points out that in a growth mindset, "people believe that their most basic abilities can be developed through dedication and hard work—brains and talent are just the starting point. This view creates a love of learning and resilience that is essential for great accomplishment."

10

The Highest Sense

Dr. Neil Theise, a New York University liver pathologist with an extraordinarily high PQ, had just seen something historic through his microscope.

He and his collaborator Dr. Diane Krause had turned off the lights in the laboratory and switched on a high-powered scope, and now they were standing in the dim fluorescent glow of it, marveling at what they saw: a slide of a liver cell from a female mouse—with a Y chromosome. Females, by definition, don't have Y chromosomes; they have two X chromosomes.

Dr. Theise was challenging the theory that adult stem cells in the bone marrow could do nothing but produce blood cells. In a previous experiment, a team had taken stem cells from the bone marrow of male mice and injected them into female mice who had had their bone marrow destroyed. Months later, these female mice had, as expected, Y chromosomes in their blood and bone marrow. What was unexpected was that these Y chromosomes were also found in other tissues, such as the lungs, stomach, and liver. How had an adult male mouse's bone marrow stem cell become a liver cell? Was this adult stem cell "plastic"?

Afterward, Dr. Theise walked home, still stunned, to his studio apartment on New Haven Green and pondered it all. *We just knocked out a foundation of Western medicine*, he thought, incredulous.

It was 1999. The stem cell field was just developing. Researchers knew that plastic stem cells could be found in babies' umbilical cords, but discovering plastic adult cells would be a watershed for medicine.

Dr. Theise didn't scream with excitement. He was still gobsmacked by it all. Though he had theorized this, to see it under the microscope in the ruby and emerald and sapphire stains used to show the cellular details was another thing altogether.

Everything I've taught my students isn't true! he thought. *Is there even such a thing as "cells," or are they simply groups of molecules able to reassemble? What is the nature of things?*[1]

His apartment had no furniture except for a mattress on the floor, a meditation cushion, and a small table for his computer. In fact, it looked just like a different type of cell: the austere space of a monk or a prisoner. Dr. Theise sat there in the darkness looking at the night sky through enormous windows reaching up to fourteen-foot ceilings. The only light was the blue of his computer monitor. He grabbed a box of Cocoa Krispies and a container of milk and polished off bowl after bowl, thinking, *What would Einstein do?*

Einstein might try a thought experiment, as he had when he was developing his theory of relativity—he'd imagined what would happen if he rode alongside a beam of light.

"Einstein's first great thought experiment came when he was about 16," wrote Walter Isaacson in a 2015 *New York Times* essay marking the hundredth anniversary of the theory of general relativity. "Einstein tried to picture what it would be like to travel so fast that you caught up with a light beam. If he rode alongside it, he later wrote, 'I should observe such a beam of light as an electromagnetic field at rest.'"[2]

Neil Theise imagined he was sitting on a cell as it moved from the bone marrow through the blood to the liver, and a fast-moving vision unfolded in his mind's eye. As the cell began to travel in the body, the scene looked animated, Dr. Theise told me. The cell was not glowing brightly like the one under the microscope back at the lab; it was more naturally and subtly colored in tones of gray.

"Whatever I was defining as a cell, I now understood was not solid but made of some smaller stuff that was changing over. And frankly speaking, I knew it was at the level of molecules and atoms, but I wasn't visualizing molecules and atoms. I was visualizing a scene where things were flying off the cell as in a wind stream and things were coming in and it was reassembling." The cell transformed before his eyes. All that remained of the original was the double helix of DNA.

A photo of Dr. Theise's mouse slide ended up on the cover of the journal *Hepatology*. By 2001, he had proven that the adult mouse stem cells could differentiate into many other types of cells, and he was featured in the journal *Cell* in May of that year. By August, his stem-cell work and that of about a dozen other researchers led to an unprecedented eleven-minute cautionary address to the nation by President George W. Bush. It became a political football—some stem-cell research involves fetal tissue from human embryos—and funding grew hard to get.

Dr. Theise is a synesthete and compassionate, and I believe this is instrumental in both his imaginative sense and his creativity in science. Recently, he helped discover a new organ—the human interstitium (layers of the body, once thought to be solid connective tissue, that are actually interconnected fluid-filled spaces)—using similar imaginative reveries. He is writing a book about complexity theory, demonstrating that not only are cells made of self-organizing molecules but that the whole universe is one vast self-organizing system.

Dr. Theise has practiced Zen Buddhism for over thirty years, using it to hone his inner sight and mental acuity. This discipline, I believe, has raised his imaginative sense to virtuoso levels.

Dr. Theise's professional activities have helped improve his inner visual senses. "When I found the field, what felt good about pathology is I loved looking at the slides. I loved being in that space. When I'm looking through the microscope, I'm in a different world, I'm in the world of the slide. I don't see the room around me."

The ability to see into the imaginative realm is the highest sense of all, and in my model of PQ, it propels one into the category of genius. Not many people can do it as well as Dr. Theise; some people cannot do it at all.

Recently, he told me that after he gave a lecture on his experiences to a small group of academics, a woman in the audience approached him. She told him she had no capacity to visualize, not even while reading a vivid novel. It was just words on a page to her and thoughts in her head. He found that remarkable.

She'd had no idea that people saw things when they read until she'd heard his talk, he told me. And he'd had no idea that there were people who couldn't. This condition— known as *aphantasia*—has only recently been documented. According to the BBC, as many as one in fifty people have no mental imagery.

In 2011, the extraordinary savant and synesthete Daniel

Tammet spoke to me about where numbers come from. He sees moving colorful forms for his math and noted that "whole schools of mathematicians believe numbers come from another dimension." He likened that place, which helped him memorize pi to 22,514 decimal points, to a landscape.

"Different numbers have an identity, and those identities are obviously the result of the way that they relate to other numbers," Tammet said.[3]

Einstein may have been a synesthete himself. Consider this quote: "If I were not a physicist, I would probably be a musician. I often think in music. I live my daydreams in music. I see my life in terms of music. I get most joy in life out of music."

Isaacson called for a celebration of "the visualized 'thought experiments' that were the navigation lights guiding Einstein to his brilliant creation."

These thought experiments remind us, Isaacson said, that "creativity is based on imagination. If we hope to inspire kids to love science, we need to do more than drill them in math and memorized formulas. We should stimulate their minds' eyes as well. Even let them daydream."

Dr. Theise used to think of his discoveries as setting off cascades of dominoes falling forward and away from him into the distance. Now, he says, he sees himself on a boat sailing on a river, and there are walls that seem to be limits across the river, but they keep dissolving.

He still does frequent thought experiments, and the space in which he travels is the same kind of space in which he experiences time as a synesthete, as he told me in a previous interview: "open, unbounded, freely rotated."[4]

"Perception and imagination are linked because the brain uses the same neural circuits for both functions," Gregory Berns, professor of neuroeconomics and director of the Center for Neuropolicy at Emory University, told *Fast Company*. "Imagination is like running perception in reverse."

Imagination uses both the executive and default networks in the brain. "When you're remembering, thinking about the future or imagining alternative scenarios," the *Fast Company* article continued, "you're activating the default network; what cognitive psychologist Scott Barry Kaufman of the University of Pennsylvania's Imagination Institute calls the 'imagination network.'"[5]

The best way to fire up your imagination is to do novel things, according to the scientists. In this way, your brain won't be relying on stored information from experience. "Seek out environments you have no experience with," Berns told the reporter. "Novel experiences are so effective at unleashing the imagination because they force the perceptual system out of categorization, the tendency of the brain to take shortcuts."

Imagination may hold promise for treating post-traumatic stress disorder, according to researchers at Colorado University. In a study published in the journal *Neuron*, researchers found that "imagination is a neurological reality that can impact our brains and bodies in ways that matter for our wellbeing," according to Tor Wager, director of Colorado University's Cognitive and Affective Neuroscience Laboratory.[6]

In that study, participants were trained to associate a sound with an uncomfortable (but not dangerous) electric shock. They were then split into three groups. One group had the sound repeatedly played for them (without being shocked); the second group was asked to play the sound in their heads as best as they could; and the third group (essentially a control group) was asked to think of pleasant sounds (birds singing and rain falling). The participants' brain activity in all these phases was measured by fMRI scans and skin sensors.

The group that had been asked to think of pleasant nature sounds continued to show the same fear response when the sound was later played for them. But researchers found that those who'd actually heard the sound (without the accompanying shock) and those who'd only imagined the sound (also, of course, without the accompanying shock) both experienced extinction (meaning that eventually, they stopped being afraid of the sound). Using only the imagination, the brain had "unlearned to be afraid."

"The idea of 'crisis management' requires no explanation right now," wrote Martin Reeves and Jack Fuller in the *Harvard Business Review* in 2020.[7] "Something unexpected and significant happens, and our first instincts are to defend against—and later to understand and manage—the disturbance to the status quo. The crisis is an unpredictable enemy to be tamed for the purpose of restoring normality."

But what if we can't return to the pre-crisis reality? The authors believe that imagination—"the capacity to create, evolve, and exploit mental models of things or situations that don't yet exist"   is the way to find new paths to growth.

"With imagination," they said, "we can do better than merely adapting to a new environment—we can thrive by shaping it. To do this, we need to strategize across multiple timescales, each requiring a different style of thinking."

They proposed seven strategies for creating more imaginative work environments, but I believe these ideas can be applied on the individual level as well:

- Carve out time for reflection.
- Ask active, open questions.
- Allow yourself to be playful.
- Set up a system for sharing ideas.
- Seek out the anomalous and unexpected.
- Encourage experimentation.
- Stay hopeful.

Stephen T. Asma, professor of philosophy at Columbia College Chicago and author of *The Evolution of Imagination*, and actor Paul Giamatti teamed up to explore the imaginative sense in an inspired essay for *Aeon*.[8] They pointed out that in *The Descent of Man*, Charles Darwin wrote: "The imagination is one of the highest prerogatives of man. By this faculty he unites, independently of the will, former images and ideas, and thus creates brilliant and novel results. . . . Dreaming gives us the best notion of this power."

But not all philosophers are impressed with the power of imagination. "For Plato, the imagination produces only illusion," Asma and Giamatti wrote. "By contrast, Aristotle saw imagination as a necessary ingredient to knowledge. Memory is a repository of images and events, but imagination (phantasia) calls up, unites and combines those memories into tools for judgment and decision-making." Aristotle saw imagination as an almost involuntary process, the mind trying to make sense of one's experience. But if we can bring the process of imagination under executive control, they said, and harness it to the conscious creative mind, "we transform the involuntary imagination to voluntary, and this 'phantasia 2.0' is unique to humans. Perhaps a chimpanzee might dream of a hippo it once saw, but only a Walt Disney can bring the hippo to mind whenever he wants, dress it in a tutu in his mind's eye, draw it, animate it dancing, and release it as a film called *Fantasia*."

Academic institutions are now exploring the power of imagination. The University of California at San Diego's Arthur C. Clarke Center for Human Imagination operates under the belief that "every innovation, revolution, paradigm shift, and step in social progress was imagined before it was enacted." The center explores the neuroscience of imagination and creates sensory-enhancing and virtual-reality technology. Philadelphia's Imagination Institute "is dedicated to making progress on the measurement, growth, and improvement of imagination across all sectors of society." Programs like these will no doubt raise the perception quotient of humanity going forward.

Earlier, I introduced you to Lesley Roy, a super-smeller who caught the scent of her own cancer long before she had symptoms, in chapter 1. She had some imaginative insights while waiting to hear back from her doctors. Once again, art was the conduit for sensitivity.

"I started a painting called *Intuition*. It was on the easel, paint still wet, now finally finished on January 7, the day of the phone call with the cancer diagnosis. I remember after the call I cried for an hour or two, called my mom and best friends, then walked over to the painting and couldn't help but notice the powerful resemblance and the realization that the most critical information that came to me while painting this may have saved my life."[9]

A friend had come over during the holidays and shared some passages from her favorite book, *Women Who Run with the Wolves* by Clarissa Pinkola Estés, the Jungian psychologist. It would inspire Roy's composition.

"The myths and stories of the wild-woman archetype have long informed my feminine psyche. I had never painted a portrait before, but this blond woman with an upswept do, wearing a lush, collared fur coat, looked just like me . . . Noticeably flared nostrils appeared to breathe in deeply, nose-to-nose with the wild wolf, as if to suck in the life-giving primeval air for a starving soul . . . Her calmly closed eyes were seeing without sight. Insight! Her pursed lips were exhaling a puff of frosty air that melded into the wolf's chin. They were becoming as one." Her golden necklace with its petite, pea-size nugget held the real secret—a small tumor.

Roy painted her ear prominently as a focal point, bringing the sense of hearing and listening to the center. It was grounded by a natural pigment, raw umber, derived from the soil of the earth. She wanted it to look a little dirty, as if she had just rubbed her ear in the earth to mark her scent—a way to find her way home in the dark.

Roy painted a white highlight along the jugular vein from ear to clavicle. Later, she noticed that it mirrored the exact incision line of the surgery.

"In the process of shutting down my logical left brain, had I activated my creative right side of the brain and let a flood of information come in? I've known that creativity is the lifeblood of the soul and has played a major role in healing in my life before. Could my creativity be informing me of a greater danger that lies ahead? Perhaps I knew I had cancer all along—the information was there, I just wasn't seeing it, feeling it, knowing it to be true," she said. "As I painted over the weeks and coming month, it occurred [to me] that we all have a vast potential to know, to know what cannot be explained." The golden-nugget necklace in the painting was in the same spot that her cancer was found. Had she been listening so acutely to her body that she was painting what her subconscious already knew?

"Was I transforming/awakening with each brushstroke? Has our modern world diluted our senses? Can we reawaken them?"

11

The Eyes of Eve

Even before I learned I was a tetrachromat, I suspected that there was something peculiar about the way I saw the world. Not only did I enjoy color more than most people, but I had many polite disagreements about the hues of things.

"Look at this carrot! It's so red!" said a friend who'd just come back from the farmers' market.

"Really?" I said. "To me, it's beet-colored. It's more like purple."

"Voilà! Baby blond!" my hairdresser said as she swung me around in the salon chair to face the mirror. *Too yellow, too brassy, yikes!* I thought, and I asked her to tone it down.

"Here are some more beige paint samples for the exterior of your house," said a patient contractor. I went through them and gently pronounced this one too orange, that one too yellow, this other one too gray, and that one not blue enough—undertones that were completely lost on him as he stood there wearing an exasperated look that seemed to say, *Lady, it's just beige.*

(*Just* beige? Heresy!)

But I got it—I felt frustrated too. I couldn't understand why people didn't see colors the way I did or, for that matter, why I didn't see the colors they saw. Their descriptions were always a few clicks to the right or left of where I believed shades fell on the spectrum, and sometimes they were just too general. What was just *red* or *yellow* or *blue* to them was *scarlet* or *citrine* or *indigo* to me.

Society's color wheel looked anemic. And things I read and descriptions I heard didn't ring true to me either. For example, how could people talk about "red" cardinals when they were vermilion with orange tones? Was it a language thing, and was I just a color-vocabulary nerd? I wondered. Sunrises I photographed were never as colorful in the resulting pictures as they had been to my eyes. Even when the images were beautiful, they often left me flat. More than that, the spectrum of colors I saw in various mediums—on TV, on the computer screen, and in print materials—was far narrower than the palette of colors I saw in the natural world.

My routine eye exams never turned up anything unusual. I thought perhaps it was my synesthesias, which often provide many color impressions. But none of my fellow synesthetes seemed to be struggling with this.

Synesthetes are in fact better than non-synesthetes at remembering colors. I participated in a study through the University of Missouri at St. Louis that seemed to connect color memory with synesthesia. In the study, computer monitors showed a photograph of the sea and a shoreline, and subjects were asked to stare at it for a minute. The next screen was completely white, and we were asked to consider the saturation level of the afterimages we saw there, choosing from six different photographs, each at a different level of focus and color intensity. The study showed that synesthetes, including myself, had better color memory than non-synesthetes.

In 2013, I stumbled on a *Radiolab* podcast[1] that began to help me unravel the mystery. In the report, the hosts discussed a trait called *tetrachromacy* and interviewed Susan Hogan, a talented interior designer who had the trait and said it enhanced her color vision. The experiences she related—from seeing tones of pink and red in an otherwise clear blue sky to her general enthusiasm for the color spectrum to, yes, the feeling that she saw color differently—were so familiar, I felt like they could have been interviewing me.

When the hosts visited Hogan in her home and noticd d she had a jukebox, they asked her to play "A Whiter Shade of Pale" by Procol Harum, and she did. I still get emotional thinking about it.

I phoned the vision expert featured on the show, Jay Neitz of the Neitz Lab at the University of Washington in Seattle, to see if he thought I might be a tetrachromat.[2]

Neitz asked me several questions about my and my family's history, including whether any of my male relatives were color-blind. That seemed like an odd query—weren't we talking about enhanced visual ability? But I did have several male relatives with color blindness, and when I told Neitz that, it seemed to be the clincher.

"That's it! I'm sending a spit kit!" he said.

I later came to understand that tetrachromacy is related to color blindness. Humans have three types of cones (the structures that allow you to see color). The genes for two of those cones—the L-cone, which responds to colors in the red spectrum, and the M-cone, which responds to colors in the green spectrum—are located on the X chromosome. (The third type of cone, the S-cone, which responds to the blue spectrum, is located on an autosome—a chromosome that's not related to gender.) Women have two X chromosomes; men have an X and a Y chromosome. If the gene for, say, the L-cone, which allows you to see colors in the red spectrum, is mutated so its sensitivity is shifted toward the yellow/green part of the spectrum, a man won't be able to see red very well. Since women have two X chromosomes, a mutated L-cone gene on one X chromosome won't affect a woman's color vision because the other X chromosome with the genes for normal color vision take over.

Only one X chromosome is activated in a cell at a time,

and the choice of which X chromosome is turned off is random. A woman who has a mutant L-cone on one X chromosome and a normal L-cone on the other will have a fifty-fifty mix of L-cone types in her retinal cells, meaning that she will have four types of cones: the normal S-cone, the normal M-cone, the normal L-cone, and the mutated L-cone. This, theoretically, allows her to see more colors of the spectrum than people who have just three types of cones. In reality, this isn't always the case—some people are genetic tetrachromats (meaning that they have the genes for four types of cones), but functional trichromats (meaning that their brains don't use the extra visual input, so they have typical color vision). Functional tetrachromats are rare.

In effect, for families with this genetic anomaly, nature robs some of the color spectrum from the men and gives it to the women.

Dr. Neitz sent me a DNA test along with written color exams. As soon as the package arrived, I set to work on the questions. Some of the images looked like the Ishihara test for color blindness—groups of dots of various hues that form a pattern that's visible only if you're not color-blind. I found Dr. Neitz's tests slightly subtler and more challenging than the ones at the optometrist's office, however. As directed, the next day, I woke up early and collected an uncomfortably large amount of saliva for the DNA kit before even having a glass of water. Then I mailed it off.

It was a very long six weeks as I waited for the results. Though I was almost sure the tests would show I was a tetrachromat, the podcast had emphasized how exceedingly rare the trait is.

Finally, an email arrived: "I just wanted to let you know that you do have the genetic basis for four types of cone photoreceptors," Neitz wrote. And his lab director, Toni Haun, followed up, informing me that I'd scored 100 percent on each of two written tests. This suggested that I might have both the genotype and the phenotype for tetrachromacy. (Your genotype is your genetic code; your phenotype is how your genes are expressed. For instance, you might carry the genes for both blue and brown eyes, but only one of those genes will be expressed in the color of your eyes. Genetic tetrachromats are not necessarily functional tetrachromats, although it turned out that I am.)

Normal people can see one million colors; tetrachromats have the potential to see a hundred million. I was a super–color seer.

"Way to *know thyself*!" said one friend.

"You're a unicorn!" said another.

"That's some X-Men stuff right there," said a third. "It's all fun and games until adamantium claws shoot out of your hands," he warned.

Suddenly, what had always been strange was sort of cool. I finally had an answer. I felt relief. And I was excited at the thought of embarking on a completely new personal journey.

I already knew a bit about my genetic profile and my small but significant place among the world's people because I had joined the Human Genome Project (a gift from my parents one year; they'd ordered the kit for me). The test showed I belonged to haplogroup H—a genetic lineage common to people of Northern European descent. But my

mutations—one way of identifying early humans' migra-
tion patterns around the world—indicated that my genetic
heritage went much further back in time: my people were
in the second big group to leave East Africa.

The scientists could pinpoint through these genetic mark-
ers where my ancestors had gone after that. They had traveled
up through the Saudi Peninsula into Central Asia, then up to
what is now Russia, and then down into Northern Europe. A
map with red lines showing the routes was included in the
results package, and it never ceases to fill me with wonder.
After the tetrachromacy results, I wanted to know even more.
Several commercial kits I've done show that I have twice as
many Neanderthal genes as almost all the other participants
in the various pools. I don't know if this has anything to do
with my hyper-sensory nature, though I suspect it may.

Neitz explained that color vision was linked to our primate ancestors, and I realized this story was much older
than my forebears in Germany's Neander Valley. Were it
not for a single female Old World monkey's mutation for
full-color vision millions of years earlier, humans would
still see on a gray scale—if we had even evolved at all, he
said. He said that the vast diversification of primates and
the resulting evolution of human beings did not begin until
monkeys developed a full color spectrum.

It is possible, he added, that the evolution of trichromacy might have been a key step in the evolution of the
primate and, ultimately, the human brain. "With the ability
to discriminate food with a higher nutritional value using
color vision, it may have been an evolutionary advantage to
expand the cognitive capacity of the brain to analyze the

new sensory signals. Acquiring color vision may, thus, have been an important contributor to human brain evolution."

Would humans have survived if they didn't have the ability to choose unspoiled fruit and see predators at a distance? It was a huge step. We might owe our existence to the eyes of an Old World monkey Eve.

Neitz said tetrachromacy might also be an evolutionary leap. "The development of trichromatic vision in the ancestors of modern primates may have been key to the evolution of humankind, and tetrachromacy may be another leap," he told me. He foresees a day when it spreads to the entire population.

Alan R. Templeton, a geneticist and statistician at Washington University, wrote that the recent increase in human population means an increase in mutations in the human gene pool, which enhances the potential for human evolution. "The increase in human population size coupled with our increased capacity to move across the globe has induced a rapid and ongoing evolutionary shift in how genetic variation is distributed within and among local human populations."[3]

I am currently being studied by a group of researchers at Arizona State University. The team is led by Kristopher Jake Patten and includes Nobel Prize winner Frank Wilczek. I, along with two other women I recommended to the team—Susan Hogan and her fellow interior designer Megan Arquette of Los Angeles—have been through two rounds of innovative testing so far, with very positive results for functionality. We can see things a control group of non-tetrachromats cannot. The research is continuing.

Patten said the goal is to create a widely available, inexpensive test that bypasses the DNA test currently in use and goes straight to behavioral function. "The current test we have for identifying tetrachromacy is genetic. Simply put, that test is expensive, poses privacy concerns, and, while genetically accurate, is not always behaviorally accurate. The gene for tetrachromacy may exist but not be expressed. Or it may be expressed but not in a sufficient quantity of cones to make a perceptual difference."

It's possible that some people do have the cones expressed but, because almost all the nonnatural colors in our world are trichromatic, they don't notice the information the fourth type of cone is picking up, he explained. "Genetic testing can identify nonfunctional tetrachromats. My colleagues and I are working toward making a reliable, cheap, and easy-to-take behavioral test for tetrachromacy."[4]

When people ask me what it's like to have this trait, I joke that it's like living the "What Color Is the Dress?" internet controversy every day of my life.

On February 26, 2015, a washed-out photo of a dress a Scottish woman had worn to her daughter's wedding broke the internet. Caitlin McNeill posted a picture of the dress on Tumblr with the caption "Guys, please help me—is this dress white and gold, or blue and black? Me and my friends can't agree, and we are freaking the f—— out."

While it seemed everyone in the world had an opinion of what color the dress was—even Taylor Swift tweeted about it—in the end, as I wrote in *Psychology Today*,[5] "It really matters what tetrachromats say." Interestingly, while normal viewers debated the colors, two of my fellow super-seers and I essentially agreed on the tones—none of which were exactly the black and blue, blue and gold, white and gold, or blue and brown being discussed. As I observed, "The dress is a fascinating retinal Rorschach test, but one must consider the limitations of trichromatic, or 'normal,' vision."

What color was the dress for me? I wrote, "It is periwinkle and bronze in the light shown, though I would like to see it in person and in natural light to be certain." So I was on Team Light Blue and Brown, but I was dissatisfied with the lack of specificity in the shades of blue and brown offered.

"It's all about the lighting!" said super-seer Susan Hogan. "In brighter light, I see copper-gold and pale lavender-white. If you tilt the laptop screen back [reducing the amount of light], the dress appears blue and black to me."

Megan Arquette agreed. "I can't judge the dress unless I see it in front of me. The image isn't the actual dress. The image reads periwinkle and bronze. By the way, snow, even when one sees it in person, looks blue or purple in the shade."

I was thrilled to see people get so passionate about colors. It gave me a cultural touchstone. Perhaps now I could talk with non-tetrachromats about the subjectivity of color vision!

Not long after that, someone on LinkedIn posted a pop test for tetrachromacy that went viral. Though such tests may be suggestive of the trait, the only valid way to prove this gift is through testing one's genes. And even then, further testing must be done to prove the person has functional tetrachromacy.

I am one of just a handful of tetrachromats identified

in the world today. We are always women—or butterflies, goldfish, birds, or, in the past, dinosaurs.

How did we get this way? The theory is that one must have the genes as well as an exposure to color and its related vocabulary to be a functional tetrachromat.

As a child, I delighted in all I could see and was particularly drawn to color. In kindergarten, I fell in love with paints and construction paper and, especially, crayons. I could occupy myself for hours independently and had a very rich inner life with Crayola's burnt sienna and maize and cornflower and beyond. I loved the crayons' names almost as much as the hues. But I often blended the colors, because even with sixty-four in the box, there were so many more I could see. (They don't blend well, but I tried.) In 1972, when they released eight new fluorescent shades, I was beside myself.

Color was and remains the furniture in my memory palace. In addition to the synesthesia pairings I already mentioned, there is the soft teal that reminds me of my mother's embrace because of a sweater in that shade she used to wear; the deep orange of June's tiger lilies reminds me of my paternal grandfather because they pleased him so; the gleam of brass that always reminds me of polishing my father's police-uniform buttons; the lilac that evokes taking the Regents Examinations because the flowers always bloom in May when they are given; the seafoam of my favorite childhood pajamas . . . I may not remember what I ate for lunch yesterday, but were you to ask me to paint these shades from long ago, I know I could match them exactly.

✳

We can all learn to see colors we haven't seen before. K. Anders Ericsson conducted a famous experiment with a female subject known only as A. F. in which he taught her to see colors previously imperceptible to her.

In *The Exceptional Brain: Neuropsychology of Talent and Special Abilities,* Ericsson explained that people's ability to identify and name colors "is often assumed to be far superior to other perceptual identification skills . . . But such experience is deceiving." The number of hues that humans could identify correctly, according to a 1951 study, might be as low as three or as high as ten. But could someone learn more through practice?

Every day for more than a year, the researchers presented A. F. with twenty-one similarly colored paint chips from the Farnsworth-Munsell 100 Hue Color Vision test, in which colors change so gradually that when they are lined up correctly, there seem to be no boundaries between them. They found the most improvement in her perception when she verbalized each color's qualities. For example, when she said one day that chip number sixteen appeared bright, she was able to identify it correctly in thirty-three of the next thirty-five trials.[6]

I learned more about color's power in high school when I was chosen to represent my New York City borough at Empire Girls State, an educational political workshop (William J. Clinton attended its brother organization for young men when he was a teenager). During my week upstate in Cazenovia, I was excited to meet many other young women from across New York and learn the inner workings of our government, and I came to understand how important color is in politics. Though I learned a lot there, perhaps it's no surprise that when I reflect on that formative experience, the thing I remember best is this: pale green is a good choice for candidates' clothing because it is viewed by people as nonthreatening and calming. That is why it is also often used to paint classroom walls.

And we learned what a disaster Vice President Richard Nixon's gray suit was against the gray background in his infamous debate against John F. Kennedy. The first-ever televised presidential debate took place on September 12, 1960. Kennedy was a young Catholic senator from Massachusetts and the underdog by far going into the event. Nixon was an experienced statesman, but he looked sickly and lethargic from a recent hospitalization. Most analysts talk about Nixon's energy level and the fact that he was sweating as the main factors behind his disastrous showing. But it was also the fact that on black-and-white television screens, he was nearly indistinguishable from the studio's walls. He quite literally faded into oblivion (for the time being), while Kennedy, in a dark suit, stood out. Learning that was a watershed moment for me, since before, I had only sensed that color conveyed emotion and power.

There is a theory that superabilities result not just from innate gifts but also from obsession. I may have been given an extra dose of color perception at conception, but there's no doubt that I spent a lot of time honing my skills, although I was not aware I was doing so. In my late teens and twenties, I used to love to string together the lavishly hued glass beads my parents, then working as antiques dealers, bought in box lots at auctions. I still have a container full of these gorgeous vintage Czechoslovakian bijoux. I was exacting in finding just the right shades for my creations, delicately placing them on white paper or holding them up to sunlight. I was a little precious about it, spending what observers seemed to think was an inordinate amount of time considering each piece. One necklace I created had tones graduating from lemon to honey to deep amber, all lined up according to their subtle color differences. I took great pleasure in finding the perfect shades for this golden rainbow, even though I'm not particularly fond of yellow. For me to enjoy that color, it needs to be cool and soft like a pastel, or at least without the orange undertones of school buses and New York City taxicabs, which are practically intolerable to me.

I've been a photographer since college; I spend time studying images, and I'm very particular about my own. In May of 2000, I was granted permission from the Archaeological Survey of India to photograph the Taj Mahal at dawn, before the thousands of tourists were let in. It took three hours of pleading with bureaucrats over many cups of tea in a steamy, ramshackle office under a squeaky ceiling fan to get that request approved. I wanted that fabled monument to love to stand alone in my pictures, with no other forms or their colors to detract from the stark white wonder of the world.

Guards met me and my driver in darkness and opened the gate. I set up my tripod in front of the reflecting pool and changed the factory lens on the camera to a special fish-eye lens that would create two 180-degree panoramic shots when I swung the device around manually. (Software would later weave these two hemispheres together so that people could mouse over the entire landscape when viewing it online.) I captured the historic shot without a soul to be found on the grounds, but there was something that eclipsed even that privilege for me.

The sun came up over the fabled tomb of Mumtaz Mahal and shone through the marble, creating a pearlescent translucence I had never previously witnessed. The glow lit up the walls from within, and they sparkled with iridescent shimmer. When I tell that story now, I am still conflicted about which is the bigger news, the unprecedented 360-degree photo or the pearlescent translucence.

After I completed my shots and packed up my gear, I encountered a busload of Indian tourists arriving in Agra to see their nation's greatest monument for the first time. I deduced it was their first time because of the looks of utter wonderment on their faces. As they were milling around, an older woman in a deep pink sari stopped in front of me. (The great *Vogue* editor Diana Vreeland said, "Pink is the navy blue of India.") Her silk garment was a color I'd admired on crape myrtle trees, and there were so many other shades of rose present in the group's attire. She pointed at my eyes and said something to her friends that was lost on me; however, they all turned to me and stared, startled looks on their faces. We stood there for a moment, entranced by each other.

From the time I was an infant, my eyes have always preceded the rest of me, foreshadowing my destiny. As a baby, I looked like a figure in a Margaret Keane painting, with outlandishly large peepers floating like two full moons above my fat cheeks. My eyes are still big and translucent and a little bit spooky; as Edgar Allan Poe said, there is no beauty without some strangeness in the proportion. People stop me on the street to ask if my eyes are naturally that pale blue—a clear, sparkly, and watery-looking aqua—or if I'm wearing colored contact lenses. In a feature article in 2014, *Vogue* called me the "Girl with Kaleidoscope Eyes," and in 2016, I was the inspiration for and color consultant on M·A·C Cosmetics' Liptensity, a bestselling line of super-pigmented lipsticks.

As I matured, I came to respect my eyes for what they could show me rather than how they looked. They gazed upon the Alhambra in Spain, the Eiffel Tower in Paris; they watched candles set aboard tiny ferries twinkle under the great minarets of Istanbul on the Bosporus one night. They read beautiful prose; they survived seeing the second plane crash into the World Trade Center. And now they might be useful for science, which is the most gratifying part of all.

Kristopher Jake Patten, the researcher at Arizona State University, theorizes that I might be able to spot COVID, or at least the beginnings of an illness, on someone's face from the flushing of the skin caused by fever. He believes that tetrachromacy evolved in part because it allowed a tetrachromat woman to spot early signs of a fever in a baby. She could also spot changes in another person's skin tone, giving her insight into people's emotions. This could help her to succeed socially and avoid danger, as fluctuations in blood flow to the face are often telling. A person might "blanch" with fear or become reddened with rage, and a tetrachromat woman could easily see it earlier than most other people.

There is precedent for using the abilities of those with vision differences in times of emergency. People with color blindness were used in World War II and Vietnam to spot enemy camouflage, Dr. Patten told me. A few recent studies show that people can distinguish between ill and well faces, possibly based on coloring. He thinks that tetrachromats might be even more sensitive to this coloration difference.

While he hasn't tested this theory yet, in the age of COVID, I find myself scanning faces in crowds even more than I did before. I see many different tones, and I also feel things from mirror-touch synesthesia.

Once, someone dear to me developed a very pale, small, pink discoloration on his fair-skinned face. The mark appeared to have a different texture than the rest of his skin; it was smoother and shinier. I insisted he see a doctor. The spot was diagnosed as precancerous; he was treated and is healthy today possibly because I was so adamant. Now I understand why evolution favored the trait of tetrachromacy—it can give one a survival advantage.

Leading neuroscientist Dr. Joe Tsien told me that the trait is advantageous. Tsien developed what is known as the *theory of connectivity,* and he studies the relationships among behaviors, genes, and neural circuits. A major evolutionary advantage to tetrachromacy is social recognition, he explained to me. People who have the trait are better at detecting hints of angers and displeasure based on subtle changes in facial blood flow.[7]

There may be other applications for tetrachromacy. For instance, I know when storms are coming by the slightest changes in the sky. The heavens are fluid to me, and the colors are water-based—a small addition of a different hue changes them completely. And there are many colors throughout the day apart from the blazes of dawn and sunset. In *The Book Thief*, Markus Zusak wrote, "People observe the colors of a day only at its beginnings and its ends, but to me, it's quite clear that a day merges through a multitude of shades and intonations, with each passing moment. A single *hour* can consist of thousands of different colors. Waxy yellows, cloud-spat blues. Murky darknesses."[8]

One morning, I saw the clear light blue dome above me turn a few drops toward green, and I urgently warned a group of people I was standing with to get inside before tornado winds picked up. An awful storm erupted.

We are only beginning to think about the applications of this newly discovered sensory ability.

Long before the first tetrachromat woman was documented—in 2010 in England—H. L. de Vries, a Dutch scientist, theorized she existed. In 1948, De Vries tested the vision of color-blind men—men with two normal types of cones and one mutated type of cone that made them less sensitive to, depending on the mutation, either red or green light. He had them mix red and green lights on a lab instrument to make what looked to their eyes like a match of a particular shade of yellow. To compensate for their difficulty in discerning hues, color-blind men needed to add more green or red than normal trichromats to make a match, according to an article in *Discover* magazine. Out of curiosity, De Vries then tested the daughters of one of his subjects. He found that even though they were not color-blind, they needed more red tones in their light than normal controls did to feel they had perfectly matched the yellow.

What was that about? De Vries wondered.

He theorized that since color blindness ran in families, women might also be affected, but in another way. The M- and L-cones are on the X chromosome, so if a red/green color-blind man had the genes for a normal M-cone and a mutated L-cone, his daughters, who inherited one X chromosome from him, would have one X chromosome with a normal M-cone and a mutated L-cone and one X chromosome (from their mother) with a normal M-cone and a normal L-cone. They would therefore have the genes for three normal cones (S, M, and L) and one mutant cone (the abnormal L-cone). He suspected that fourth type of cone was why the female relatives of the color-blind subjects perceived color differently—not because they saw *less* than most people but because they saw *more*. He speculated on the last page of his paper on the topic that such women might distinguish more colors than a typical trichromat, but he didn't pursue the idea.

In the 1980s, Cambridge neuroscientist John Mollon, who was studying color vision in monkeys, became interested in De Vries's brief mention of tetrachromacy. He and one of his grad students, Gabriele Jordan, theorized that since color blindness was fairly common, tetrachromacy might not be that rare.

Jordan and Mollon tested the mothers of color-blind sons using a color-matching test similar to the one De Vries had used. They hypothesized that a tetrachromat would not be able to make what she considered a perfect match because she would see far more color gradations than those available for her to use.

But that wasn't what they found. All the mothers of color-blind sons—who, given the laws of genetics, must have had the genes for four types of cones—had no problem making what they considered perfect matches. So maybe they were wrong, the scientists thought; perhaps there wasn't such a thing as functional tetrachromacy.

In 2007, Jordan, then a neuroscientist at Newcastle University, tried a different method. She had the subjects sit in a dark room and look into a lab device that periodically flashed what appeared to be three orange circles in rapid succession; for each trial, the subject pushed a button to indicate which circle, if any, looked different from the other two. "To a trichromat," an article in *Discover* reported, "they all looked the same. To a tetrachromat, though, one would stand out." That circle was not pure orange but a subtle mixture of red and green. "Only a tetrachromat would be able to perceive the difference, thanks to the extra shades made visible by her fourth cone."[9]

Jordan tested twenty-five women who had four types of cones. One of those women, code-named cDa29, chose the correct circle every time in repeated trials and got perfect scores in several other experiments. "I was jumping up and down," Jordan told *Discover.* They had at last found their tetrachromat.[10]

Jordan, now a senior lecturer at Newcastle University's Centre for Transformative Neuroscience, reported on the first genetically proven human tetrachromat in the *Journal of Vision* in July 2010. Jordan has since found a few other tetrachromats, but "it is extremely difficult to identify tetrachromats 'behaviorally,'" she told me during an interview for *Vogue*. "There are many more individuals who have the genetic disposition than there are individuals who use it perceptually."

In July 2012, *Discover* magazine published one of the first articles about these tetrachromats. It pointed out what has been called "the paradox of tetrachromacy": How could four-coned women know they saw extra colors when they lived in a world made by trichromats?

"Most of the things that we see as colored are manufactured by people who are trying to make colors that work for trichromats," Jay Neitz noted. "It could be that our whole world is tuned to the world of the trichromat."

And how does cDa29 see the world? "She was unable to communicate her experience to the researchers in much the same way as it is impossible to describe the experience of red to a dichromatic person," Jordan told *Discover*.

I believe there are more hues in nature than there are in human-made objects. Mother Nature is a tetrachromat, at the very least, if not a pentachromat or even a hexachromat. Gazing on my backyard trees, I see shades of green that I've never seen anywhere else. For example, in bright sunlight, the leaves have a white-silver-green glow that is nothing like anything made by humans.

While I see many beautiful hues in nature that others don't, I also notice a lot of muddy, unattractive tones, colors I have never seen in manufactured goods—for example, there's a certain "meh" autumn-leaf tone that would never be used for a sofa textile. My home is filled with things made by trichromats for trichromats, but when I step out my front door and into nature, I feel like Dorothy opening the door into Oz sometimes, except, unlike garish Technicolor, the bright and beautiful colors I see have subtle undertones.

I've begun to log my tetrachromatic colors and name them. The odd, muddy-green brown (not quite khaki) at the center of one of the glass bits on my desk lamp is *plicon*. The mind-blowing color in a sunset that is both baby pink and champagne without turning peach is *jeresha*. That unattractive leaf color I mentioned is *breem*.

I do not believe trichromatic human-made colors are as finite as perceived. Unless a designer has the neuro-equivalent of an internal paint-mixing machine—where precise amounts of blue, yellow, black, white, or brown are released to come up with the perfect shade desired—there are going to be plenty of unintentional colors. When I see a piece of fabric that is a little off from the ideal of beauty or current trends, I think the trichromatic creator arrived at an "accidental" color. I don't believe trichromats see the variation, but I do.

Recently, a color believed to be a new shade of blue, now known as "YInMn Blue," was accidentally created in an Oregon State University lab by a group of chemists developing new electronics materials from manganese oxide. According to *Tech Times*, the team "erroneously mixed black manganese oxide with various chemicals and heated the mixture to about 2,000 degrees Fahrenheit. Surprisingly, one of the samples came out a bit different than what they expected: an intense, impossibly blue pigment."[11] The shade—which, incidentally, didn't strike me as new but as something I've seen in well-lit lapis lazuli stones—has since been licensed by the Ohio-based Shepherd Color Company.

We knew about tetrachromacy in the animal kingdom long before we found it in humans. Paleontologists discovered not long ago that most dinosaurs had feathers—and this was millions of years before the first reptiles started to fly. Many of the fossils turning up in the world, from Siberia to South America, exhibit evidence of plumage. German scientists at the University of Bonn have theorized that this is because dinosaurs were tetrachromats and could better see the feathers for display and mating purposes. "Until now, the evolution of feathers was mainly considered to be an adaptation related to flight or to warm-bloodedness," the study's main author, Marie-Claire Koschowitz, said in an interview with the *Huffington Post.* "I was never really convinced by any of these theories."

Koschowitz and her team analyzed genetic similarities between dinosaurs and modern-day birds and reptiles. "If you look at a cladogram"—a diagram that illustrates evolutionary relationships—"for a group of animals for which the relationships are well known and you find a feature that is shared by all of them, it's pretty safe to assume that this feature was present at the base of the tree and kept throughout the evolution of the last common ancestor into the different species," Koschowitz said. "So I looked at the morphology and general color vision in reptiles and birds and lo and behold, it turned out that tetrachromacy is present in every single branch of today's reptiles.

"This means dinosaurs likely used visual signals to communicate with each other," Koschowitz said. Large, sheetlike feathers would produce a huge variety of colors and patterns that dinosaurs could use to recognize one another and to choose a mate.[12]

I spend alot of time photographing latter-day dinosaurs— my fellow tetrachromats, the birds. I can't resist their colorful plumage, and I take many extreme close-ups of feathers; I have an entire Instagram account devoted to the pursuit. Along the way, I noticed iridescence in birds not ordinarily credited with iridescence. Blue jays, cardinals, horned larks, rusty blackbirds, and woodpeckers all shine in the right light. After an exhaustive search of the literature revealed no reference to this, I began to collect photographic evidence and file it away. Only a minority of the hundreds of images I took showed it. I asked trichromat friends if they could see it in the pictures, and they could but they had never noticed it before on birds. Then, recently, a study on iridescence in birds came out of Princeton University; the authors explained that the iridescence was produced by crystal-like nanostructures in the birds' feathers.

I wrote the lead researcher, Klara Nordén, about my observations of iridescence in birds not generally thought to be iridescent. She wrote back:

"I think you're absolutely right—many more birds than we know are iridescent. It's always been my pipe dream to map out all iridescent birds to study how and why such beautiful colors evolve. The beauty of the colors themselves was really what got me interested in studying them in the first place."[13]

There are other modern-day creatures that signal to each other in this beautiful way. In 2008, a group of Japanese scientists discovered that tetrachromacy is present in yellow swallowtail butterflies, the first invertebrates found to exhibit the trait.[14]

Perhaps there are hues in nature beyond even what I can see with my tetrachromatic palette. Consider the color blue. Ancient languages did not have a word for blue—not Greek, Chinese, Japanese, or Hebrew. Homer, writing in *The Iliad* and *The Odyssey*, described the seas as "wine-dark," honey as "green," and sheep as "violet."

As *Radiolab* reported in its feature "Why Isn't the Sky Blue": "While black is mentioned almost two hundred times and white around one hundred by Homer, other colors are rare. Red is mentioned fewer than fifteen times, and yellow and green fewer than ten. [William Gladstone, Homer scholar and future prime minister of Great Britain] started looking at other ancient Greek texts and noticed the same thing—there was never anything described as 'blue.' The word didn't even exist."[15]

Philologist Lazarus Geiger followed up on Gladstone's work and noticed this was true around the world at the time. He reviewed an ancient Hebrew version of the Bible, Icelandic sagas, Hindu Vedic hymns, the Koran, and ancient Chinese stories and found no reference to blue. Geiger believes that since there was no name for the color blue, the brain had no category for it, and therefore, while it existed in the sky and the ocean, people were not noticing it.

There is even a debate about whether colors exist at all. The ancients believed colors were outside of us, but Galileo planted them firmly in the brain with his experiments. And in modern times, the debate has two main camps: realists who say color is what it is, out there in the environment (the ball is blue), and anti-realists who say it is nothing but wavelengths later assigned a value in our minds (your brain perceives the wavelengths reflecting from the ball as blue; if you weren't there, it would not be blue).

But there is a third point of view, and it makes the most sense to me. In her book *Outside Color: Perceptual Science and the Problem of Color in Philosophy*, University of Pittsburgh professor M. Chirimuuta writes, "Of all the properties that objects appear to have, color hovers uneasily between the subjective world of sensation and the objective world of fact." She proposed another way of looking at color: *adverbialism*. "Color is not an object of sight but a way of seeing things," Chirimuuta wrote. "Instead of a brown dog, Chirimuuta wants us to see the dog brown-ly," wrote Malcolm Harris in a *New Republic* review of her work.[16]

I checked in with David Chalmers, the philosopher of consciousness, and he asked me this poignant question: Could you explain what it's like to be a tetrachromat to someone who is not a tetrachromat?

The answer is no.

Philosophy, in particular the philosophy of mind, has long wrestled with this question. They use the term *qualia*, which means, roughly, the subjective quality of experiences, like seeing a red rose or tasting chocolate. You could know all there is to know about roses and chocolate but still not have the true knowledge of them if you didn't subjectively experience them.[17]

12

The Senses, Meditation, and Consciousness

We've already learned from several super-sensors and experts how attention and practice can help sensory development. Joy Milne, the nurse in Scotland who can smell Parkinson's disease, teaches children to pay attention and smell everything twice. Tea sommelier Marzi Pecen taught us that mindfulness is important to sensory development.

William C. Bushell reported on the virtuoso skills of adept meditation practitioners.

Studies have shown that meditation increases the ability to discern touch with a fingertip and markedly improves hearing. Further, in one study, patients with glaucoma were able to decrease their intraocular pressure (the pressure in their eyes) with meditation.

To study the effect of meditation on touch, researchers followed twenty experienced Zen Buddhist scholars on a four-day silent retreat.[1] The practitioners had an average of eleven years of training in the discipline and meditated for eight hours a day. The scientists asked half of them to be completely aware of any sensory perceptions arising in their right index fingers—a practice known as *focused-attention meditation*—for two of those eight hours every day for three days. The other half meditated as they usually did, not voluntarily focusing on any object, a practice known as *open-monitoring meditation*.

The meditators were tested on their tactile abilities before the retreat began and then on the third and fourth days of the exercise: With eyes closed, each meditator was touched on the tip of the right index finger with either a single needle or a pair of needles with an adjustable distance between them. After each touch, participants had to say whether they'd felt one needle or two. This is known as the *two-point discrimination threshold*—the smallest distance between two needles that still allows the subject to feel two separate needles. The same procedure was performed on the right middle finger and the left index finger.

Researchers found that the focused-attention meditators—the ones told to focus on the right index finger during their meditation—had "significantly enhanced tactile acuity" at the end of the retreat. The other group, the open-monitoring meditators, had no changes in their scores. "These data indicate that merely being aware without external stimulation or training can drive highly specific changes in tactile perception," the researchers wrote.

We've seen how extraordinarily sensitive our hearing is; it responds to a broader range of stimuli than any of our other senses. But the overproduction of adrenaline from stress can reduce blood circulation in the inner ear, which can cause hearing loss over time or sudden hearing loss. Studies have shown that meditation not only reduces the stress response and decreases the loss of hearing; it can actually improve hearing.

According to several studies, meditators have increased brain tissue in parts of the brain responsible for attention, reported *The Hearing Review*. "An increase in these areas would suggest that you would be better at paying attention if you meditate, but it's not just focusing. Meditating increases an individual's ability to perceive sounds at different auditory thresholds." It also improves the way the brain codes and stores auditory information.[2]

Several devastating diseases can decrease vision, but a recent study suggests that one of them, glaucoma, can be improved by mindfulness meditation. The American Academy of Ophthalmology reported that meditation may help lower eye pressure in glaucoma patients.[3]

Patients with certain kinds of glaucoma have increased

intraocular pressure, or eye pressure, which can damage the optic nerve and permanently reduce vision. If glaucoma is not treated, it can lead to total blindness. However, a study appearing in the *Journal of Glaucoma* suggested that "mindfulness meditation may play a role in helping patients cope with their disease and may actually improve outcomes."[4]

William C. Bushell said that "when adept practitioners develop the capacity to sit completely still for long periods and learn to focus attention on their sensory-perceptual processing, to reduce cognitive and attentional distractions, and allow the normally unattended precision and depth of their innate sensory capacities to emerge," they can develop a whole new sensory gestalt. *Gestalt* is, essentially, a collection of elements that together become more than the sum of their parts.[5]

Together, the senses, which meditation clearly enhances, are a big part of your consciousness. I have been seriously exploring the topic of consciousness for more than a decade. I've led synesthesia workshops at the Science of Consciousness Conference twice, in Tucson and in Stockholm. After studying for many years, I've decided I am a panpsychist, which means that I believe that consciousness permeates everything. I don't believe a brain or a system must grow complex before the lights can go on. I believe that plants and animals are parallel nations of sentience and that they, like humans, have consciousness.

I'll tell you how I came to this belief.

Bushell phoned me one day and asked if I'd like to witness a meeting between Russian dot-com billionaire Dmitry

Itskov and Tibetan lama Phakyab Rinpoche on the topic of uploading human consciousness into a robot by 2045. This may sound like science fiction, but it is not. These are not fringes, they are frontiers. And they're big business.

The meeting between the Russian mogul and a very elite monk took place in the monk's apartment. Itskov had been on the cover of the *New York Times* Business Section just days earlier and was spearheading the Global Future 2045 effort exploring such questions.[6] Rinpoche, with whom I have studied, is a master of several yoga traditions, including phowa, in which he can shift his consciousness out of his body, breaking corporeal limits. This consciousness projection is a technology—an art and a science—that is thousands of years old. There is extensive literature on the topic, although out-of-body experiences are ignored by the mainstream and considered speculative—wrongfully, in my opinion—in the West.

At the meeting, which I like to call "The Day I Saw the Fourteenth Century Advise the Twenty-First Century," Itskov and his assistant brought an iPad loaded with videos of the most state-of-the-art robots in existence. The monk sat stationary on his pillows in the living room of his Jackson Heights, Queens, apartment and watched stoically as they flicked through clips of everything from the robot in Japan who can climb stairs to Itskov's own robot, which looked so much like him, down to its sandy-blond eyelashes and blinking, lifelike hazel eyes, it was startling.

Eventually, Itskov asked Rinpoche the important question: Could human consciousness be transferred to a robot?

Rinpoche leaned in and said in a serious tone that Tibetan

monks had, in fact, perfected the transfer of consciousness many centuries ago. Consciousness could be transferred to a rock or a turtle; the consciousness of an old, sick person could be transferred to someone younger and healthier. In a deep, somber voice, he said it was a very bad idea and had been banned in the fourteenth century.

Itskov quietly put his iPad down, and before long, he and his assistant said their polite goodbyes.

Phakyab Rinpoche, a highly educated man who has been recognized by the Dalai Lama as a reincarnate lama, gave me the foundation of my panpsychism that day. He is also a master of "iron hook phowa," a beautiful practice in both Buddhism and Hinduism in which one helps transfer the consciousness of a beloved departed being (or oneself) to a better realm in the afterlife or a desirable reincarnation. Given Rinpoche's noted mastery of this tradition, I was convinced that the senses and consciousness are even more powerful than we know.

I have had the honor of knowing Indo-Tibetan scholar Robert Thurman, now retired from Columbia University, ever since his son Ganden and I tried to collaborate in housing the vast archives of Tibet House, the Manhattan cultural center that the Thurman family administrates, on a pan–South Asian website I was editing. These keepers of the dharma, or teachings of Buddhism, are His Holiness the Dalai Lama's friends and fierce defenders of his people.

I checked with Dr. Thurman—the first Tibetan monk in the West—on the latest research about the senses and asked him if he feels vindicated now that science is con-

firming what adept meditators like the Tibetans have long
reported.

"Of course, yes," he said from his home in Woodstock,
New York.[7] "The main point is that Western science is
imprisoned in the dogma of a kind of very poorly-thought-
through materialism and sort of stuck in the belief that the
sort of world that we habitually perceive is the real one."
He went on to say that that was contradicted by atomic theories and physicists' theories on the quantum. The fact that
machines are proving that the world is not only what people
perceive "is not a matter of vindicating some ancient Buddhist," he said, "it's a matter of seeing better what the nature of reality is, which should be the purpose of science."

Thurman believes that physicists and biologists should
be trained in philosophy, that they should be yogis and yogi-
nis, and that they should develop their minds by observing.

Thurman taught me that the senses and how they work
become apparent upon death, when they fade out, one after the other. And there is a supra-sense, a sense above all
others, that can lock into one or more of them at a time. In
Tasting the Universe, I wrote, "This is known in Buddhism
as the 'mind sense.' It figures in life as well as in death. Dr.
Thurman's description of this sense reminds me of a dash-
board of sorts: 'Its job is it chooses one of the senses to align
itself with or perhaps several at once. It can override and
simulate the sense organs,' he explained. He said that this
sense is active in sleep—when we dream, for example." The
dissolutions Dr. Thurman referred to are the stages of death
described in detail by Tibetan teachers, a peeling away of

the layers of what makes us conscious, living beings. The last sense, the mind sense, is what remains.[8]

Dr. Deepak Chopra has done a lot of work on the senses and consciousness. In his book *Metahuman,*[9] he addressed the importance of the senses in human potential. Dr. Chopra and I met a decade ago at the Science of Consciousness Conference and began a conversation about synesthesia that continues to this day. Dr. Chopra is working with some of the top scientists on the planet to find answers about the nature of consciousness.

Recently, he invited me on his *Conversations at the Intersection of Cutting Edge*[10] program to discuss the senses. He said that at the most fundamental level, we are ingesting only energy and information, "but we don't experience that; we experience the world, we experience our body, we experience our iPhone, or we experience a book."

The point he was trying to make in *Metahuman* was that the physical world we experience, including our physical bodies, "is a confederation of these perceptual activities. And these perceptual activities are, in turn, a modification of consciousness. So when I look at you or look at a phone or look at my hand, if I were a baby, I would not know that that's Maureen Seaberg or that is a phone, or this is my hand or that is a book. I would, in fact, be bathed in consciousness with these fluctuating particles and waves. I would be immersed in that 'goo,' so to speak, because my body is also a part of that. And somehow the process of human construction, through language, categorization, memory and culture, comes out."

That's where the hard problem of consciousness comes

in, Dr. Chopra said. How do we convert this energy and information into experience? That phrase, the "hard problem of consciousness," was coined by noted philosopher David Chalmers, a professor of philosophy and neural science at New York University and codirector of its Center for Mind, Brain, and Consciousness. In *The Character of Consciousness*, he wrote:

> It is undeniable that some organisms are subjects of experience. But the question of how it is that these systems are subjects of experience is perplexing. Why is it that when our cognitive systems engage in visual and auditory information processing, we have visual or auditory experience: the quality of deep blue, the sensation of middle C? . . . It is widely agreed that experience arises from a physical basis, but we have no good explanation of why and how it so arises. Why should physical processing give rise to a rich inner life at all? It seems objectively unreasonable that it should, and yet it does. . . . The really hard problem of consciousness is the problem of *experience*. When we think and perceive, there is a whir of information processing, but there is also a subjective aspect.[11]

"We can take our senses in at the level of quanta," Dr. Chopra said. "By *quanta* we mean the smallest indivisible units of energy that come to us in the form of photons or vibrations of the atmosphere or molecules—it doesn't matter—but at the most fundamental level, all we are ingesting is energy and information."

How do we convert these impulses of energy and

information into the experiences of colors, sensations, shapes, images, feelings, and thoughts that somehow create the constructs that we call mind, body, and world? How does that physical world produce consciousness? "That's where we get stuck," said Dr. Chopra. "It's the second most open question in science: What is the biological basis of consciousness?"

There are no colors or shapes or forms in the physical world, period, Dr. Chopra said, no sounds or fragrances or sensations or emotions; all of it is constructed in consciousness. "There's no reflection, there's no theory, there's no science in the physical world. The whole thing is constructed in consciousness."

Still, he said, we get so bamboozled by language and constructs and culture and thousands of years of conditioning "that we believe that there's a physical world and there is a physical body in that physical world when the physical world is really an ever-changing continuum of perceptual activity of consciousness and so is the body and so is the mind. . . . Ludwig Wittgenstein, the philosopher, said, 'We are asleep. Our life is a dream. But we wake up sometimes, just enough to know that we are dreaming.'"

Dr. Chopra believes that there's no such thing as a physical body. "It's a projection—a conglomeration or a federation of qualia that we interpret as the physical world and the physical body."

This reminded me of the cell discovery of our mutual friend Dr. Neil Theise. Dr. Chopra explains, "Whatever it is, the dreamscape can be infinite galaxies, or it can be a photon and you and I are fictional characters in that collective dreamscape most of the time. Yes, most of the time we agree on the dream but then the synesthetes come along and they say, 'No, I'm dreaming something else here right now.' Consciousness is not in the body nor is it outside the body. It is formless, so it has no form.... It has no boundaries, so it's infinite, but it expresses itself as a species- and culture-specific experience."

The great sages and seers and prophets all say we create our own reality. Dr. Chopra said, "And if we actually begin to understand synesthesia, we might have a solution to the two most open questions in science. One is, What is the universe made of? And my answer to that is the universe is made of consciousness. And the second is, What's the biological basis of consciousness? And man says there's none."

Dr. Kenneth R. Pelletier heralded the present time in "The Chutzpah Factor in Altered States of Consciousness," published in 1977, noting, "We have limited our growth through our own beliefs rather than through necessity. By daring to dream, imagine and challenge these self-imposed limitations, we can learn to fulfill our human potential."

Sri Sri Ravi Shankar said that with spiritual growth, "there is keenness of observation. You become totally relaxed, yet at the same time, you possess sharpness of awareness, and strength of intelligence. Your senses become so clear. You can see better, think better, hear better. Like a pure crystal, your senses come to reflect all objects as one divinity."[12]

And the Buddha said, "Train your eyes and ears; train your nose and tongue. The senses are good friends when they are trained. Train your body in deeds, train your tongue in words, train your mind in thoughts. This training will take you beyond sorrow."[13]

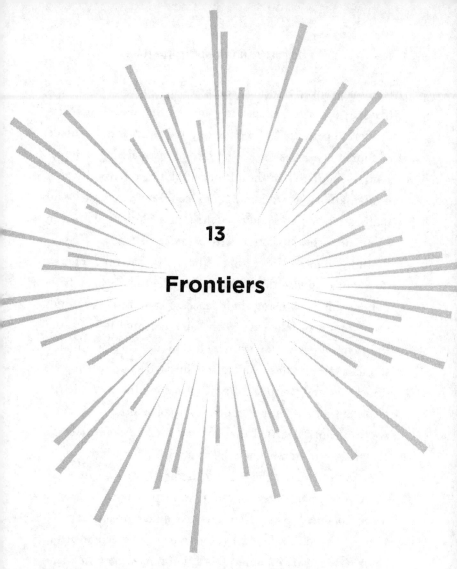

13

Frontiers

Our species stands at a crossroads. We are facing a possible transhuman future, one in which people use technology to transcend human sensory limits. We must choose wisely, not only with our brains but with all our extraordinary senses, or forever lose our humanity. I think the future is not about robots; it's about us and achieving our human potential.

The *technological singularity* is the point at which

technology overtakes and surpasses us. Sometimes it feels like we're in a rocket speeding toward that day. "The ever-accelerating progress of technology and changes in the mode of human life gives the appearance of approaching some essential singularity in the history of the race beyond which human affairs, as we know them, could not continue," said mathematician, computer scientist, and physicist John von Neumann, who coined the term *singularity* in 1950.[1]

But I don't believe that's the only dynamic. There's plenty of evidence that humans are adapting to, even passing, machines by using their senses. Just look at the mind-meld young people have with their personal devices and the manual deftness with which they use them compared to older generations. Maybe the singularity moves in two directions, and we meet somewhere along the way. There's a human-based component not yet considered. And since we've recently learned that human sensory potentials are far greater than we knew, perhaps we have a little more time to think about the value of being *Homo sapiens.*

A very small subset of the neurological outliers known as *mirror-touch synesthetes* have extreme empathy for machines. They are sometimes able to feel the mechanisms in their own sensitive bodies. I call them *machine synesthetes* or *machine empaths.*[2]

Jade McLeod of Missouri is one of them. She used to accompany her grandfather to junkyards as a child. For her grandfather, a mechanic, it was about finding parts. But for young Jade, it was an intensely emotional experience. "I remember seeing so many items, like cars, just stacked up. It felt like such a waste of life or something. As I grew older,

I started noticing that when I would go into my grandpa's shop, I would feel empathy for things like his old blue truck and for other tools he kept and allowed us to play with as children."[3]

McLeod is a rare sensory outlier, a mirror-touch synesthete with profound empathy toward people and other living things as well as machines. She and several others like her are compelling figures in our advanced technological age. So empathic are their minds and bodies that they are not just anthropomorphizing and ascribing human qualities to the technology; they are internalizing them.

"I watch a lot of movies and shows about cyborgs, androids, and AI," she told me. *Westworld, Ex Machina, Raised by Wolves, Blade Runner*, and *Love, Death & Robots* are some of her favorites. "I also like to use pieces of old technology in my art when I find them so that I feel like they have a second life." She has always been described as kindhearted and tenderhearted. "It relates to lots of things in a mechanical and technological sense as well," McLeod said. Seeing an art installation where a robotic arm worked ceaselessly made her feel intense sorrow. "I couldn't help but feel awful for the actual machine itself."

Her earliest memory of feeling bad for an appliance was a reaction to a cartoon. "I was in kindergarten, and as a reward, we were given popcorn and were allowed to watch *The Brave Little Toaster* in the library. Clearly, the toaster and other appliances are given human traits in the cartoon, which could explain why so many people could feel emotions along with the characters, but I have met several people who think it is ridiculous to empathize with a cartoon. Soon after, I saw the film *Short Circuit* [about an escaped robot and a woman who befriends it] and had the same sort of emotion." She reacted the same way to R2-D2 and C-3PO from *Star Wars*.

McLeod recalled exploring a vacant building with her brother when she was a young girl. They discovered an enormous number of abandoned computers. "I remember just staring in amazement because they had zero life. They didn't look like there was even really anything wrong with them. They looked relatively new at the time, and it honestly hurt my feelings. My brother is extremely practical and said something about the waste of money, but I did not care about the money; it was more about the abandonment for me."

McLeod finds it difficult to get rid of old and obsolete technology. "I still have CDs, tapes, records, and DVDs, and old thumb drives . . . it is just very hard for me to just throw them away."

Robin May worked as a logger in north-central Australia. Heavy machinery was essential to her job, and she felt very close to the equipment, she told me. "I owned big dozers and excavators, and I felt I could sense the emotions of the machine. I wanted to touch it to feel its journey."[4]

When she and her husband started a fence-building company, she was extremely sad about leaving the logging equipment. "I found it hard to let them go. The machines were part of my family. They, too, worked hard. They were not just machines; they were part of the entity of our logging business, which was also a vessel that carried three families, many other businesses, and connected a community. When I sold my machines, I was so happy I knew the other logger and that he was a good person. Machines matter too. My machines also had a connection with the forest. They were kind and logged with integrity and environmental care. Me and the machines felt the forest and had an extremely close relationship with our area." May told me, "I have a connection to metal chains. The feel and sound of them take me home. They feel like a connection and old friends."

She had known of her synesthesia since childhood but kept it a secret. "It wasn't until my early thirties my husband heard it on a radio show or podcast and got me to listen to it. I finally did tell some people, and a few had these experiences." But she didn't mention her empathy for machines. She and her husband own big tractors for their fencing business, and May is happiest when they are functioning well. "It's an emotion, an instinct. I feel sad when the machine is broken. I think machines have a purpose, and I feel if they are broken and not in use, they feel unworthy."

May pointed out the danger of working with heavy equipment in remote sections of the bush and believes that was a factor in her empathy. "It's extremely dangerous, and people die logging. You needed to respect the machines. We relied on each other for the safety of each other."

Michelle Peck-Harris of Vancouver, Washington, now an engineer in the wireless industry, worked with robotics when she was getting her degree. When she flipped on the switch for a robotic arm she built and it began to move, she felt sensations in her own arm and felt compelled to move in the same way.

"As it stands now, we are very far away from artificial intelligence ever beating out human intelligence in terms of what I suppose we call humanity," Peck-Harris told me.[5] "Sure, we have technology that can learn and compute data faster than any human can, but regarding having humanlike robots, I think we still lack the fundamental understanding of consciousness, considering we don't fully understand how consciousness works in humans." She believes that if humans do manage to create near-humanlike robots, their senses would outperform ours, since robot technology can constantly be updated. For instance, she said, robots could have ultraviolet or infrared sight along with normal human sight, giving them an advantage over us in visual processing.

Sensitive as a child, Peck-Harris was diagnosed a few years ago with a sensory-processing disorder. I feel this is as much a gift as a challenge for Peck-Harris. "It means that most of the time I felt too much and would get sensory overload in certain environments," she said. "Lots of signs pointed to it growing up. One example that comes to mind was that as a young child, I hated watching TV on those big old [cathode-ray tube] TVs. There was something about the sound they made and the way films were played on them." She had to cover her ears and shut her eyes whenever someone put a movie on. It was sometimes bad enough that she would hide under a chair or run out of the room. Instead of getting sympathy, she was usually scolded for being "dramatic."

"I hated loud noises in any form, even noises that weren't necessarily loud to others, such as the dishwasher running or a toilet flushing. It wasn't even just the sound; it was as if I could feel something all over my body with these sounds as well, which makes me think it was probably that mirror-touch synesthesia showing through in a different way.... I've learned over time how to cope with sensory overload, thanks to a lot of new guides and materials that have come out for people with ASD, who also typically struggle with sensory overload."

She also has a very sensitive sense of smell. "Of course, this usually meant I had to be the one to smell the milk carton in my family for spoilage.... I think synesthesia is one part having a natural sensitivity to the environment and another part being mentally in tune to the environment." She's noticed that if something dulls her senses—grief or a depressive episode—she'll "lose any form of synesthesia I had while that episode lasts."

Peck-Harris sometimes needs to enter an electromagnetic compatibility chamber, or EMC, to test a wireless device she's working on. The rooms are shielded from radiation and all forms of electromagnetic waves. "You tend to forget just how much goes on around us—both from radio frequencies that occur in nature and from those that are man-made—until you step inside one of these chambers and seal the door," she said. "It's unexplainably calm in there, and sometimes even disorienting. While we have found in our studies that the man-made radio frequencies we use in our technology have little to no physical effect on the human body, I have always wondered if we are still somehow able to sense it. We definitely sense more things in our environment than we're consciously aware of, I think. And mirror-touch synesthesia is definitely a very wireless experience! I look at something and my brain decides to tell me how to physically feel what the object or person I'm looking at is feeling—all from a distance!"

She continued, "I think we are already living in the future where nearly any body part could be easily replaced with machine parts or lab-grown tissue. It's definitely a tricky subject because on the one hand, evolving toward machines could solve a multitude of health-related problems that we face today; on the other hand, as with all modern technology, it is almost certain that this technology would be exploited."

In graduate school, Peck-Harris did research on a wireless arterial catheter that monitored a person's cardiac activity in real time, with the results easily viewed on a simple app. "Incredibly interesting stuff, considering how it could collect data that we normally wouldn't be able to collect in a regular doctor's visit."

However, wireless technology is extremely vulnerable to security hacks, even with the most robust encryption methods, Peck-Harris pointed out. "There's no saying what a person with malicious intent could do if they managed to hack into a patient's wireless implant. I think the march toward singularity is also a march toward immortality; it would greatly change our economic and social structure if humans were able to achieve immortality through robotics. I would be concerned about a social class divide widening further— when we achieve this technology, who gets it? Will it be expensive, or will it be a regular part of health care? If the lower and middle classes are unable to afford this technology while the wealthy can easily afford it, what will happen? Politics aside, though, I think the movement toward singularity is actually the natural direction for humans—humanity has always been striving to improve itself."

Like so many of the supersensitive people in this book, Peck-Harris enjoys the outdoors. "[In the forest] where we live now, it is so dark and quiet at night, and that's where I tend to find my solace. I find that I'm most productive during this time."

Recently, Peck-Harris's superpowers have been heightened in the happiest way. "I'm currently pregnant, and because of that, my sense of smell and taste have been enhanced so much," she said. "Sleep has also changed too; my dreams are so vivid now that each dream feels like a fever dream. . . . It also makes me wonder about how the body is able to enhance these senses just through hormones, and whether or not we could artificially enhance senses in the future in a similar way. If that were the case, then we wouldn't necessarily need to add something new within us to get this effect; it's already there."

Futurist Ray Kurzweil wrote in *The Singularity Is Near: When Humans Transcend Biology* that the age of virtual sentience is upon us.[6] By 2030, he said, "Nanobot technology will provide fully immersive, convincing virtual reality. Nanobots will take up positions in close physical proximity to every interneuronal connection coming from our senses. We already have the technology for electronic devices to communicate with neurons in both directions yet requiring no direct physical contact with the neurons."

※

William C. Bushell believes most of humanity hasn't properly explored or developed their senses yet. "We of the contemporary global 'cosmopolitan' culture have yet to give ourselves a chance to explore these technologically unaided sensory-perceptual capacities in ourselves," he said, "capacities which have been recognized, developed, cultivated, and apparently applied in cultures which possess adept practitioners" of particular yoga and meditational disciplines.

We are not passive observers in the nano- and quantum worlds but participants with our own amazing biotechnology to wield. That the most sensitive among us are already expressing this technological empathy is a singularity of another sort.

In *The Tao of Physics*, Fritjof Capra wrote of this interconnectivity with our environments; he believes we are more than just observers. Perhaps we are already not separate from the machines.

Capra pointed out that a Buddhist does not believe in an independent or separately existing external world into whose dynamic forces he inserts himself. Quoting Lama Anagarika Govinda, he said, "The external world and his inner world are for him only two sides of the same fabric, in which the threads of all forces and of all events, of all forms of consciousness and of their objects, are woven into an inseparable net of endless, mutually conditioned relations."[7]

Writing in the *Journal of Medical Ethics*, M. J. McNamee

and S. D. Edwards noted that both sides of the argument for and against transhumanism have "slippery slopes."[8] They recognized that medical technology has already benefited humanity but warned against some of the stronger and more dramatic technologies on the way. "We find ourselves in strange times; facts and fantasy find their way together in ethics, medicine, and philosophy journals and websites," they wrote. "Key sites of contestation include the very idea of human nature, the place of embodiment within medical ethics and, more specifically, the systematic reflections on the place of medical and other technologies in conceptions of the good life."

Stanford political scientist Francis Fukuyama called it "the world's most dangerous idea." Writing in *Foreign Policy*, Fukuyama said that while humans have many challenges, like disease, physical limitations, and short lives, as well as jealousies, violence, and other messes that make a transhuman solution appealing on one level, there are many ethical challenges.[9] "If it were technologically possible, why wouldn't we want to transcend our current species?" he asked. "The seeming reasonableness of the project, particularly when considered in small increments, is part of its danger. Society is unlikely to fall suddenly under the spell of the transhumanist worldview. But it is very possible that we will nibble at biotechnology's tempting offerings without realizing that they come at a frightful moral cost." He pointed out that such technologies will make things even more inequitable for many people across the globe.

Zoltan Istvan of California, who ran for president in 2016 and 2020 on a conservative and transhumanism platform, is, at this writing, studying ethics at the University of Cambridge to prepare for what he believes is an inevitable transition. And he told me it's all about the senses.[10] "While transhumanism is very broad in what it attempts to accomplish, achieving sentience in AI is paramount. The only way to really upgrade the human brain in singularity terms is to upload our consciousness, and once we do that, AI will be conscious," he said. "Of course, there's the issue of AI becoming sentient before we merge with it, but most transhumanists advocate against this. The reason is that the AI might not want to merge with us or it even may want to eliminate us. So we'd better get it right."

Istvan believes that, although the new research is exciting, humans will be capable of even more with technology. "We need to get rid of the perception of what is natural or common for humans. In one hundred years, having one hundred senses may be normal for our evolved species. Our brains may be connected to AI and have a million times more computational power than we have now. Once we become truly transhuman, meaning our body is more than fifty percent machine or synthetic, a new world of possibilities will await us," he said.

He thinks people are underestimating how fast AI will be here. Just because humans can't produce sentient AI today doesn't mean they won't be able to in five years (though he believes twenty-five years is more likely). "The world will be changed. This machine intelligence may transform and improve the knowledge of every science book in a matter of days. We may discover brand-new theories of the universe that truly redefine the human species. My main

concern with AI is not enough attention is being paid to it
and the ethics of its existence."

And the sensitive people I have profiled in this chapter
demonstrate something we are not yet considering as we
speed toward transhumanism.

✳

Imagine, if you will, that you open a new restaurant.
On the first night, there are nineteen very discerning
patrons. If seventeen of those elite patrons become regu-
lars for the next twenty years, your restaurant would be
a success!

This is the metaphor nuclear physicist Edwin May uses
to describe Star Gate, the remote-viewing program run by
Stanford Research Institute and various departments of the
U.S. government from 1975 to 1995. May led a team of remote viewers—that is, people who can discern information
about a distant subject using only their minds. The program
began during the Cold War in response to similar programs
in the Soviet Union.

In May's metaphor, the patrons were top government
agencies, and the menu was information that was so gourmet, the diners frequented the place for decades.

The first time I met Eddie—as he insisted I address
him—he was wearing a T-shirt from the Spy Museum that
read I WAS NOT HERE. His humor belies the deadly serious
nature of his work. He would never use the word *psychic*,
he told me.[11] To the former program leader, what the remote viewers found was *informational psi*, a serious term
he coined for very serious business. May states that his team

remotely looked in on American hostages abroad, found downed airplanes, and gathered all manner of actionable intelligence for an alphabet soup of government agencies. And they did so by utilizing not only the mind's eye but all their senses. Though the Star Gate program ended in 1995, May continues the research in his private Laboratories for Fundamental Research. "It seems unlikely that there would be such a substantial customer return rate if the data were not worthy of such attention," he said.

The International Remote Viewing Association defines *remote viewing* as a mental faculty that allows a perceiver to describe a target that is inaccessible to normal senses due to distance, time, or shielding. For example, a remote viewer might be asked to describe a location on the other side of the world, an event that happened long ago, or an object sealed in a container without being told anything about that target.

Practitioners tell me they receive smells, tastes, sounds, and sensations of touch as well as visions. Indeed, many call it remote *sensing*. This is the frontier of sensory awareness— are our senses so powerful that we don't need to be near the stimuli or even in the same era of time to perceive things?

Legion of Merit recipient and U.S. Army Special Forces veteran Joseph McMoneagle is one of the finest remote viewers this nation has ever had; he has performed more than 450 successful remote-viewing intelligence-gathering sessions. A former chief warrant officer, McMoneagle, who was injured in Vietnam, told me he has directly and indirectly advised heads of state, pharmaceutical companies, law enforcement agencies, mining companies, and seventeen of nineteen U.S. intelligence bureaus. A Star Gate veteran, he has many synesthesias, including mirror-touch. He now teaches at the Monroe Institute in Faber, Virginia.[12]

McMoneagle told me that he feels vibrations for the ages of things. "The higher the frequency of vibration I sense, the more recent something is in age. The lower the frequency, the older something might be."

The Central Intelligence Agency shut down its work with extrasensory perception in the late 1970s; the program was moved to the U.S. Army's Fort Meade in Maryland, where it was funded by the Defense Intelligence Agency. Over the better part of the next two decades, Congress continued to fund the remote-viewing program. The government denied the existence of the program, although details of its experiments leaked out in the 1980s. In 1995, the CIA released a report from the independent American Institutes for Research acknowledging that the U.S. government had attempted to use remote viewing for military and intelligence purposes.

Edwin May told me that his top three viewers during the Star Gate program, including McMoneagle, were all synesthetes and that he continues to study synesthetes in his private laboratory.

McMoneagle told me "synesthetes are shamans" and said this hyper-sensory group of traits may have saved humanity. "Synesthesia is one of the great attributes that existed even more powerfully a hundred thousand years ago. The sensitivity that you have to your environment is a survival mechanism." He believes it's why cave bears—who were twelve feet tall and weighed three thousand pounds and had the nails of nightmares—are gone, but frail humans—who are not very vicious and don't have fur or talons—are still here. Early on, human groups were units of thirty or so people. McMoneagle believes these small tribes were led by synesthetes who were what people today would call shamans. These leaders were known for their knowledge of and relationship with their surroundings, and their PQ might have helped their tribes survive.

Unfortunately, in modern times, McMoneagle said, "we believe we no longer need the shamans. But is that true?"

14

Vivify Your Senses

We live in extraordinary times. We know more about the senses than any other generation before us, but will we do the work to meet our potential and raise our PQ? Or will we just rely on brain chips and other transhuman devices?

Most people are deluded by a "tyranny of ordinary appearances," to use a tantric term, regarding the senses, which operate far beyond the parameters we've been taught. Lama

Thubten Yeshe wrote in his *Introduction to the Tantra* that such self-limiting ideas are the root of much unhappiness and stunted growth, and therefore, "if we truly wish to achieve the satisfaction of complete self-fulfillment, we must find a way to break free from the tyranny of ordinary appearances and conceptions. We must gain a heartfelt appreciation of how disastrous it is to continue relating to our body and our mind, and therefore our self-image, in the gross and limiting ways we do now."[1]

After surveying much of the new research and collecting first-person testimony from extraordinary sensors and scientists, I have created a ten-step model to help you develop better perceptual abilities and expand your perception quotient so you can free yourself from such limited thinking. It is organized by a handy and appropriate ten-letter acronym, PERCEPTION, which stands for

- Practice
- Expect
- Respect
- Chutzpah
- Escape
- Pair
- Think
- Improve
- Om
- Nature

I know there are many virtuoso sensors in our world waiting to emerge in this new climate; they just need the tools. Most people are probably already experiencing things that they are not consciously aware of or that they dismiss.

Eden Phillpotts wrote in his book *A Shadow Passes,* "The universe is full of magical things patiently waiting for our wits to grow sharper."[2] I hope we do grow sharper and see things with fresh eyes through the latest research.

I see this as the path forward to vivifying your senses.

P
PRACTICE

Practicing sensory perception will make you a better sensor.

The senses are plastic and can be refined and grown. This takes practice. We saw how super-smeller Joy Milne smells each scent twice, and we learned how a Monell Center scientist began to smell a scent he previously could not after repeated exposure to it. Sensory expert Kristopher Jake Patten advises playing a musical instrument to improve hearing.

One fun exercise I like (and all of this should be enjoyable!) is listening to a piece of orchestral music and trying to tease out the separate instruments. Can you hear the difference between a wooden flute and a metal one, for instance? Check online performances on YouTube and other platforms to learn the answer. The same thing can be done to discern the ingredients in a delicious sauce in a restaurant. You and your dining companions can ask the chef for the recipe at the end of a meal.

Throw a sensory party. Blindfold guests and present

them with various flowers. Make your own savory sauce and ask your visitors to identify the ingredients during your dinner conversation. Walk into a florist's shop or a garden center and smell deeply and mindfully while noting the names of the flowers and herbs.

E
EXPECT

Know that you are capable of much more than you believe.

The very act of expecting more from your sensorium opens you to experience. Use the new research findings to raise your personal expectations.

This step in our sensory growth is about mindset. We learned from Carol Dweck's work at Stanford University that people who have the expectation of success—a mindset of success—achieve more. You can be limited by your expectations or empowered by them.

Set a sensory goal and have confidence you'll meet it. For example, take all the perfumes and colognes in your home, close your eyes, and have someone spray each one so you can identify them. Do not worry about failure; expect that you will get many of the answers right and that you will improve with each try. If you have children, have them join you. It doesn't have to be perfume; the exercise can be done with soap or scented candles or spices.

Do you know the odors of the people you love? Laundry

is less mundane if you close your eyes, pick a shirt randomly out of the pile, and try to determine whose it is.

Take a walk in the woods, and instead of well-worn paths, allow your feet to follow your nose to the things you smell. You'll be amazed at your successes, such as sensing the musk of a deer who recently passed through.

R

RESPECT

Make the world safe for sensitivity.

Emotional sensitivity and the senses are intrinsically linked in the brain.

Respect your own impressions and those of others. Avoid people who diminish your sensations or deny you the right to have them. And don't be that person yourself.

Perhaps one of the best ways to communicate respect is to notice sensory gifts and then acknowledge them verbally and with your actions. One of my college friends was a great sounding board for the themes in this book. As I got into the olfactory and gustatory ideas, he realized his own teenage daughter, a picky eater, smells her food before eating it. He now acknowledges this behavior may be a superability and not strange, possibly an indicator of a future as a "nose" in the perfume industry.

Who in your life would be good to talk to about these things? I'm certain you can think of several. Please explore it with them gently, kindly, and with the wonder it deserves, and I will do the same.

C
CHUTZPAH

Have some.

Wield your senses with confidence.

We saw in Kenneth Pelletier's classic paper "The Chutzpah Factor in Altered States of Consciousness" that exceptional human performances have one thing in common—belief on the part of the subject that he or she can succeed in the task.

Chutzpah is different from just expecting more from your senses. Chutzpah makes you get out of your chair and assert your exquisite equipment in various environments. While senses are for the sensitive, once you respect yourself and others, you will create the space to really test these new potentials and have fun with them.

You can be sensitive and still approach life with zest and not cleave to old ideas about yourself. This extends to intuition and imagination, which are also senses. Have a hunch? Act on it.

E
ESCAPE

Practice sensory self-care. Protect your space
and avoid the constant bombardment of stimuli
modern society throws at you.

This sensory step may seem like common sense, yet how many of us practice it?

There are desirable sensory experiences and harmful ones. Steve Roach taught us to protect our ears and not become the frog in the boiling pot of water of noise over time.

Drop out of the cacophony when you can. Take a sensory-health day, like a mental-health day. Go for a walk somewhere quiet. Turn off the lights when necessary and breathe deeply.

Neil Theise is one of the most productive people I know, and he takes weeklong silent retreats a couple of times a year with his Zen community. The finest technology on the planet—the human sensorium—deserves rest and maintenance.

P

PAIR

Make bonus associations for the senses like the latent synesthete you are.

If you smell magnolia, pair it with a visual image, like a full moon or some other memorable thing. This will help you remember it better.

We are all synesthetes. But some people get the signal before it hits the screen, as we learned from the work of Richard Cytowic, and experience it consciously. Synesthetes have automatic pairings of stimuli and senses, and due to that, they remember experiences and objects longer and have better access to them.

Start with your favorite sensation, whether that's the smell of peonies, the sound of a lark, or the taste of raspberries. Make a list of sensations you would like to remember longer and more vividly, then make a list of objects or colors or music next to it and connect them all. Memorize the associations. Or, less formally, do it as they come up in your experience.

T
THINK

Be more deliberate with your senses.

Use your sensoria consciously, and be more alert in each possible moment of sense. Stop several times daily and think about what you're sensing. I call this a "sense stop."

A good time to do a sense stop is when you are eating. Most people can't remember what they ate for lunch because they weren't thinking about the blessing and nourishment of the food as they were eating it; they were worrying about their next appointments or something that happened the day before.

Sense stops also help you enjoy the now of life. The present is all we have. The senses are a great reminder of that truth. "Stop and smell the roses" is said so often that it's lost its meaning, but please reconsider it.

A great tool for this step is to write about your senses, even just a brief daily note on your calendar about a particularly beautiful sensory experience. It should be a joy, not a chore, so don't make it hard.

I promise that when you look back on a month of such notes, you'll feel happier and more alive.

I

IMPROVE

Improve your sensory vocabulary. Learn more
words for colors, flavors, and sensory stimuli
in general. Language opens the pathways to
greater functionality.

We know that superabilities and the ability to name what
the sense is perceiving are related. In my case, if I did not
have a large vocabulary for color, my tetrachromacy genes
might not have become functional.

Create pathways for your senses by expanding your related vocabulary. One way to begin building sensory flu-
ency is to start with color. Instead of just *blue,* consider the
variations—*azure, cobalt, periwinkle,* and *teal,* for example.
Look them up and then try to find them in your environment. "Cook good food," suggested olfactory expert Luca
Turin.[3] Try recipes with ingredients you're not familiar
with and learn what they are.

Try lots of things you haven't tried before in all sensory
ways. Shake it up and wake it up.

O

OM

Meditate.

Meditators reported vivification of the senses and experiences with quantum light and other phenomena centuries
before scientists discovered the upper end of our perception.

We've also learned that meditation and yoga have enormous benefits to health in general, and they can lower stress hormones that damage sensory apparatuses. In some cases, meditation can reverse damage.

Here is a very specific and appropriate meditation for the senses I recommend; it's from Lama Thubten Yeshe's *Introduction to Tantra: The Transformation of Desire.*

In tantra we focus our attention upon such an archetypal image and identify with it in order to arouse the deepest, most profound aspects of our being and bring them into our present reality. It is a simple truth that if we identify ourselves as being fundamentally pure, strong, and capable we will actually develop these qualities, but if we continue to think of ourselves as dull and foolish, that is what we will become. To give an example of tantric transformation, we might visualize ourselves as Manjushri, a princely deity usually represented as reddish yellow in color, holding the sword of discriminating awareness in his right hand and a text of Buddha's Perfection of Wisdom teachings in his left. . . . The purpose of seeing ourselves in the form of this particular deity is to hasten the development of the wisdom already within us. . . .

The more closely we identify with such a deity, having become familiar with what each of his attributes stands for, the more deeply we stimulate in our own mind the growth of the qualities he represents. . . . We will be able to perceive this self-generated deity with a clarity far exceeding that of our present self-image. Our mind will actually become the mind of the deity, and our ordinary sensory experiences—what we see, hear, taste, and so forth—will be transformed into the blissful enjoyments of the deity. This is not a fairy tale. Such transformation has been the experience of countless tantric meditators of the past and there is no reason why, if we exert sufficient effort, we should not experience the same transcendental results.

N

NATURE

Rewild yourself. Get out into nature as much as possible.

We pity the caged bird, the shelter dog, and the chained elephant in the circus, but we can't see our own sensory incarceration, much of which is self-imposed.

Humans lived outside for far longer than they have lived indoors, and people do their exquisite "soft-tissue/high-tech" bodies a disservice bordering on human rights abuse by holing themselves up inside. Researchers found that Malaysian hunter-gatherers who lived off the land had a far more evolved sense of smell than more settled people. And they were also able to identify more colors. And while many of us in the "developed" world live lives very different from that setting, the benefits of being outdoors should inspire us to do so as much as possible.

We are animals. And not just animals but super-predators—we reign supreme over all of nature. Every bit, if not more, of the instinct and sensory excellence we ascribe to the rest of the animal kingdom lives in us. Somewhere on the path to modern civilization, we forgot this truth. We are feral in all the wrong ways. Perhaps if we integrated and expanded our sensory impulses in positive ways, humanity wouldn't be so broken.

Please join me on a journey to an exciting new future in human potential. Step into the power of your mighty senses.

Acknowledgments

I'm grateful to many incredible people for their help and inspiration. First, I would like to thank my literary agent, Andrea Somberg, for believing in this work in its earliest stages and for encouraging me to ask the what-if questions about a world in which the senses are marshaled to improve our lives. You are as congenial as you are brilliant, and I feel so fortunate to have you shepherding this project.

Thank you to Elisabeth Dyssegaard, the book's extraordinary editor. I am grateful that the visionary woman who gave the world *Smilla's Sense of Snow* and who clearly understands the power of perception brought such remarkable eyes to the work.

To Tracy Roe—medical doctor, copyeditor, and so much more. You are a fantastic collaborator and just the "right hand" I needed. I have unending admiration for your clarity of mind!

Sincere thanks are also due to Daniela Rapp, who first acquired this book for St. Martin's Press and is wished every blessing in her new endeavors. You will always be the work's "fairy godmother" and inspired the last, service chapter with clear steps to vivify the senses.

To my parents, Richard and Mary Seaberg, thank you for your love and for having so many good books in our humble NYPD Emergency Service Unit Truck One household when I was growing up.

Dr. Neil Theise, you are a huge blessing in my life, providing warm friendship, inspiration, teaching, and course corrections when necessary. I can't believe I lived in the time of David Bowie *and* Neil Theise.

Kristopher Jake Patten, Ph.D., you are strikingly eloquent as both a writer and a scientist. Thank you for proving the functionality of my tetrachromacy and for teaching me so much about the senses. I can never repay what you've done for me and the many yet-to-be-discovered tetrachromat women to come.

To the many sensory outliers, especially the synesthetes, who have shared their stories here and elsewhere for years, thank you for your community. The youngest of you taught me that things once called "normal" should be thought of as merely "expected."

Thank you to Drs. Larry Marks and Richard Cytowic for having the chutzpah to research synesthesia and prove it was real. Thank you to Dr. V. S. Ramachandran for championing nature's exceptions, like me, and for your poetry. Dr. David Eagleman, you are a wonder to watch as you go bravely into the sensory future. Thank you to my favorite "cool hero" Dr. Robert Thurman for teaching me not only Buddhist metaphysics but human rights. Dr. Deepak Chopra—you are the hardest-working man in consciousness studies, and I thank you for your teachings and warmth.

To the readers I haven't met yet who will have their own sensory stories to share—I can't wait for the conversation to begin.

Bibliography

A., Nikki. "YInMn Blue: Rare Pigment That 'Absorbs Radiation' Is Selling at $179." *Tech Times*, January 25, 2021. https://www.techtimes.com/articles/256295/20210125/yinmn-blue-rare-pigment-absorbs-radiation-selling-179.htm.

Abram, David. *The Spell of the Sensuous: Perception and Language in a More-Than-Human World.* New York: Vintage, 2017.

Ackerman, Diane. *A Natural History of the Senses.* New York: Knopf, 2011.

Ambinder, Michael S., et al. "Human Four-Dimensional Spatial Intuition in Virtual Reality." *Psychonomic Bulletin & Review* 16, no. 5 (2009): 818–23. https://doi.org/10.3758/pbr.16.5.818.

Asma, Stephen T. *The Evolution of Imagination.* Chicago: University of Chicago Press, 2017.

Atwood, Margaret. *The Blind Assassin.* Boston: Little, Brown, 2019.

"Autism-Related Behaviors Are Shaped by Neurons Outside of the Brain." *Science.* https://www.science.org/content/page /autism-related-behaviors-are-shaped-neurons-outside -brain.

Ayres, A. Jean. *Sensory Integration and the Child.* Torrance, CA: Western Psychological Services, 1979.

"Bearing Witness to Mirror Touch Synesthesia." *Sensorium* (blog). *Psychology Today.* https://www.psychologytoday.com /us/blog/sensorium/201805/bearing-witness-mirror-touch -synesthesia.

"Better Odor Recognition in Odour-Color Synesthesia." *ScienceDaily,* August 21, 2017. https://www.sciencedaily.com /releases/2017/08/170821105523.htm.

Bitterman, Y., et al. "Ultra-Fine Frequency Tuning Revealed in Single Neurons of Human Auditory Cortex." *Nature* 451, no. 7175 (2008): 197–201. https://doi.org/10.1038/nature06476.

Bolognini, Nadia, et al. "Touch to See: Neuropsychological Evidence of a Sensory Mirror System for Touch." *Cerebral Cortex* 22, no. 9 (2011): 2055–64. https://doi.org/10.1093/cercor /bhr283.

Braaten, Ellen B., and Dennis Norman. "Intelligence (IQ) Testing." *Pediatrics in Review,* November 1, 2006. https://publications .aap.org/pediatricsinreview/article-abstract/27/11/403 /34094/Intelligence-IQ-Testing?autologincheck=redirected %3FnfToken.

"The Brain Detects Disease in Others Even Before It Breaks Out." *ScienceDaily,* May 24, 2017. https://www.sciencedaily .com/releases/2017/05/170524084650.htm.

Brazeau, Martin D., and Per E. Ahlberg. "Tetrapod-Like Middle Ear Architecture in a Devonian Fish." *Nature* 439, no. 7074 (2006): 318–21. https://doi.org/10.1038/nature04196.

Brower, Tracy. "Empathy Is the Most Important Leadership Skill According to Research." *Forbes*, January 12, 2022. https://www.forbes.com/sites/tracybrower/2021/09/19/empathy-is-the-most-important-leadership-skill-according-to-research/?sh=5adad3ed3dc5.

Bryson, Bill. *A Short History of Nearly Everything*. New York: Random House, 2016.

Buck, Carl Darling. *A Dictionary of Selected Synonyms in the Principal Indo-European Languages*. Chicago: University of Chicago Press, 1949.

Bushell, W. C. "Can Long-Term Training in Highly Focused Forms of Observation Potentially Influence Performance in Terms of the Observer Model in Physics? Consideration of Adepts of Observation Meditation Practice." *Cosmos and History* 12, no. 2 (2016): 31–43.

Bushell, W. C. "Neuroscientific and Quantum Physical Approach to Advanced Buddhist Mindfulness Meditation: Perceptual Learning, Neuroplasticity, Complexity, Texture, Fractals, and Synesthesia. A Model In-Progress." Towards a Science of Consciousness, 2011, Aula Magna Hall, Stockholm, Sweden.

Bushell, W. C. "New Beginnings: Evidence That the Meditational Regimen Can Lead to Optimization of Perception, Attention, Cognition, and Other Functions." *Annals of the New York Academy of Sciences* 1172 (2009): 348–61.

Bushell, W. C. "Yogic Perception: An Integrative Scientific Model of Vajrayāna Meditational Practices." *Proceedings of the Second International Conference on Vajrayāna Buddhism*. Thimphu, Bhutan: Centre of Bhutan Studies, 2018, 130–55.

Bushell, W. C., and Maureen Seaberg. "Experiments Suggest Humans Can Directly Observe the Quantum." *Sensorium* (blog). *Psychology Today*, 2018–2019.

Bushell, W. C., et al. "Human Quantum Perception: Potentially Revolutionary Historical and Future Possibilities." In progress.

Canlon, B., T. Theorell, and D. Hasson. "Associations Between Stress and Hearing Problems in Humans." *Hearing Research*. Accessed August 24, 2022. https://pubmed.ncbi.nlm.nih.gov /22982334/.

Capra, Fritjof. *The Tao of Physics: An Exploration of the Parallels Between Modern Physics and Eastern Mysticism.* Boston, MA: Shambhala Publications, 2010.

Case-Smith, Jane, and Teresa Bryan. "The Effects of Occupational Therapy with Sensory Integration Emphasis on Preschool-Age Children with Autism." *American Journal of Occupational Therapy* 53, no. 5 (1999): 489–97. https://doi.org/10 .5014/ajot.53.5.489.

Chalmers, David. "Facing up to the Problem of Consciousness." http://www.consc.net/papers/facing.pdf.

Chirimuuta, M. *Outside Color: Perceptual Science and the Puzzle of Color in Philosophy.* Cambridge, MA: MIT Press, 2017.

Chopra, Deepak. *Metahuman.* New York: Random House, 2020.

Classen, Constance. *Aroma: The Cultural History of Smell.* London: Routledge, 1994.

Cytowic, Richard. "Synesthesia: Phenomenology and Neuropsychology: A Review of Current Knowledge." *Psyche: An Interdisciplinary Journal of Research On Consciousness*, 1995.

Cytowic, Richard E. "Theories of Synesthesia: A Review and a New Proposal." *Springer Series in Neuropsychology* (1989): 61–90. https://doi.org/10.1007/978-1-4612-3542-2_3.

Cytowic, Richard, and David Eagleman. *Wednesday Is Indigo*

Blue. Discovering the Brain of Synesthesia. Cambridge, MA: MIT Press, 2011.

Dada, Tanuj, and Meghal Gagrani. "Mindfulness Meditation Can Benefit Glaucoma Patients." *Journal of Current Glaucoma Practice,* 2019. https://www.ncbi.nlm.nih.gov/pmc /articles/PMC6710928/.

"Daniel's Eyes of China Blue." *Sensorium* (blog). *Psychology Today.* https://www.psychologytoday.com/us/blog/sensorium /201205/daniels-eyes-china-blue.

"Dr. Neil Theise's Magnificent Time Wheels." *Sensorium* (blog). *Psychology Today.* https://www.psychologytoday.com/us /blog/sensorium/201202/dr-neil-theises-magnificent-time -wheels.

Duffy, Patricia Lynne. *Blue Cats and Chartreuse Kittens: How Synesthetes Color Their Worlds.* New York: Times Books, 2001.

Dweck, Carol S. *Mindset.* London: Robinson, 2017.

Easwaran, Eknath, trans. *The Dhammapada,* Easwaran's Classics of Indian Spirituality Book 3. Tomales, CA: Nilgiri Press, 2007.

Ericsson, Anders. *The Cambridge Handbook of Expertise and Expert Performance.* Cambridge, England: Cambridge University Press, 2006.

"Extraordinary Sensorium of Brian Wilson." *Sensorium* (blog). *Psychology Today.* https://www.psychologytoday.com/us /blog/sensorium/201706/the-extraordinary-sensorium -brian-wilson.

Fukuyama, Francis. "Transhumanism." *Foreign Policy* 144 (2004): 42. https://doi.org/10.2307/4152980.

Gardner, Amanda. "Love Salt? You Might Be a 'Supertaster.'" CNN, June 16, 2010. https://www.cnn.com/2010/HEALTH /06/16/salt.taste/index.html.

"Geoffrey Rush on His Synesthesia." *Sensorium* (blog). *Psychology Today*. https://www.psychologytoday.com/us/blog/sensorium/201408/geoffrey-rush-his-synesthesia.

George, Nancie, et al. "Ten Incredible Facts About Your Sense of Smell." *Everyday Health*, October 3, 2014. https://www.everydayhealth.com/news/incredible-facts-about-your-sense-smell/.

Gilbert, Avery N. *What the Nose Knows: The Science of Scent in Everyday Life*. Fort Collins, CO: Synesthetics, 2014.

Gilbert, Chris. "Musky Smell and Parkinson's Disease." https://drchrisgilbert.com/2018/02/23/musky-smell-and-parkinsons-disease/.

Goleman, Daniel. *Emotional Intelligence*. New York: Bantam, 2020.

Good, Arla, et al. "Compensatory Plasticity in the Deaf Brain: Effects on Perception of Music." *Brain Sciences* 4, no. 4 (2014): 560–74. https://doi.org/10.3390/brainsci4040560.

Greenwood, Veronique. "The Humans with Super Human Vision." *Discover*, May 24, 2020. https://www.discovermagazine.com/mind/the-humans-with-super-human-vision.

Gutek, Gerald Lee. *The Montessori Method: The Origins of an Educational Innovation*. Lanham, MD: Rowman & Littlefield, 2004.

Hall, Julie L. "When Narcissists and Enablers Say You're Too Sensitive." *Sensorium* (blog). *Psychology Today*. https://www.psychologytoday.com/us/blog/the-narcissist-in-your-life/202102/when-narcissists-and-enablers-say-youre-too-sensitive.

Harris, Anna. "Sensory School." *A Sensory Education*. London: Routledge, 2020. 65–82. https://doi.org/10.4324/9781003084341-5.

Harris, Malcolm. "Does Color Even Exist?" *New Republic.* https://newrepublic.com/article/121843/philosophy-color-perception.

Hasson, Dan, et al. "Stress and Prevalence of Hearing Problems in the Swedish Working Population—BMC Public Health." BioMed Central. February 23, 2011. https://bmcpublichealth.biomedcentral.com/articles/10.1186/1471-2458-11-130.

Hawks, John. "Where Is Human Evolution Heading?" *US News & World Report.* https://www.usnews.com/science/articles/2008/07/24/where-is-human-evolution-heading.

"He's Got a Way About Him." *Sensorium* (blog). *Psychology Today.* https://www.psychologytoday.com/us/blog/sensorium/201205/hes-got-way-about-him.

"Hilma af Klint: Synesthete." *Sensorium* (blog). *Psychology Today.* https://www.psychologytoday.com/us/blog/sensorium/201903/hilma-af-klint-synesthete.

Holderness, Cates. "What Colors Are This Dress?" *BuzzFeed,* July 28, 2021. https://www.buzzfeed.com/catesish/help-am-i-going-insane-its-definitely-blue.

Holmes, Rebecca. "An Eye on Experiments That Make Quantum Mechanics Visible: *Aeon* Essays." *Aeon,* https://aeon.co/essays/an-eye-on-experiments-that-make-quantum-mechanics-visible.

"Human Eye Can Detect a Single Photon, Study Finds." *Los Angeles Times,* July 19, 2016. https://www.latimes.com/science/sciencenow/la-sci-sn-human-eye-photon-20160719-snap-story.html.

"Imagination Is the Sixth Sense. Be Careful How You Use It: *Aeon* Essays." *Aeon,* March 2, 2022. https://aeon.co/essays/imagination-is-the-sixth-sense-be-careful-how-you-use-it.

"Is There Such a Thing as a Super Hearer?" Australian Broadcasting Corporation, July 25, 2012. https://www.abc.net.au/science/articles/2012/07/25/3553426.htm.

Isaacson, Walter. "The Light-Beam Rider." *The New York Times,* October 30, 2015. https://www.nytimes.com/2015/11/01/opinion/sunday/the-light-beam-rider.html.

Istvan, Zoltan. *The Transhumanist Wager.* Brookings, OR: Futurity Imagine Media, 2013.

Jensen, Arthur R. "Galton's Legacy to Research on Intelligence." *Journal of Biosocial Science* 34, no. 2 (2002): 145–72. https://doi.org/10.1017/s0021932002001451.

Jordan, G., et al. "The Dimensionality of Color Vision in Carriers of Anomalous Trichromacy." *Journal of Vision* 10, no. 8 (2010): 12. https://doi.org/10.1167/10.8.12.

Jraissati, Yasmina. "Reporting Color Experience in Grapheme-Color Synesthesia." Oxford Scholarship Online, 2017. https://doi.org/10.1093/oso/9780199688289.003.0005.

Kant, Immanuel. *Critique of Pure Reason.* 1787; Aegitas, 2016. Kindle.

Keller, Helen. *Helen Keller: The Story of My Life.* New York: Signet Classic, 2002.

Keller, Helen. "Three Days to See." *Atlantic Monthly,* January 1933. American Foundation for the Blind. https://www.afb.org/about-afb/history/helen-keller/books-essays-speeches/senses/three-days-see-published-atlantic.

Keltner, Dacher. *Dacher Keltner on Touch. Greater Good.* https://greatergood.berkeley.edu/video/item/dacher_keltner_on_touch.

Koshitaka, Hisaharu, et al. "Tetrachromacy in a Butterfly That Has Eight Varieties of Spectral Receptors." *Proceedings of the Royal Society B: Biological Sciences* 275, no. 1637 (2008): 947–54. https://doi.org/10.1098/rspb.2007.1614.

Kosko, Bart. *Noise.* New York: Viking, 2006.

Kuang, Shenbing. "Toward a Unified Social Motor Cognition Theory of Understanding Mirror-Touch Synaesthesia." *Frontiers*, January 1, 2016. https://www.frontiersin.org/articles/10.3389/fnhum.2016.00246/full.

Kurzweil, Ray. *The Singularity Is Near: When Humans Transcend Biology.* London: Duckworth, 2013.

LeMasurier, Meredith, and Peter G. Gillespie. "Hair-Cell Mechanotransduction and Cochlear Amplification." *Neuron* 48, no. 3 (2005): 403–15. https://doi.org/10.1016/j.neuron.2005.10.017.

Lerner, Evan. "Penn Research Helps Show That Attention, Imagination Equally Important for Creativity." *Penn Today*, August 27, 1970. https://penntoday.upenn.edu/news/penn-research-helps-show-attention-imagination-equally-important-creativity.

Linden, David J. *Touch: The Science of Hand, Heart, and Mind.* New York: Penguin, 2016.

Lipomi, Darren. "(Invited) Organic Haptics: Soft Materials for Artificial Touch." *ECS Meeting Abstracts*, no. 26 (2018): 1540. https://doi.org/10.1149/ma2018-01/26/154.

MacGregor, Andrew J., et al. "Co-Occurrence of Hearing Loss and Posttraumatic Stress Disorder among Injured Military Personnel: A Retrospective Study." *BMC Public Health* 20, no. 1 (2020). https://doi.org/10.1186/s12889-020-08999-6.

Mailer, Norman. *Marilyn Monroe.* New York: Taschen, 2017.

Majid, Asifa, and Nicole Kruspe. "Hunter-Gatherer Olfaction Is Special." *Current Biology* 28, no. 3 (2018). https://doi.org/10.1016/j.cub.2017.12.014.

Maßberg, Désirée, and Hanns Hatt. "Human Olfactory Receptors: Novel Cellular Functions Outside of the Nose." *Physiological Reviews* 98, no. 3 (2018): 1739–63. https://doi.org/10.1152/physrev.00013.2017.

Maurer, Daphne, et al. "Synesthesia in Infants and Very Young Children." Oxford Handbooks Online, 2013. https://doi.org/10.1093/oxfordhb/9780199603329.013.0003.

McGann, John P. "Poor Human Olfaction Is a Nineteenth Century Myth." *Science* 356, no. 6338 (May 2017). https://www.ncbi.nlm.nih.gov/pmc/articles/PMC5512720/.

McNamee, M. J., and S. D. Edwards. "Transhumanism, Medical Technology and Slippery Slopes." *Journal of Medical Ethics* 32, no. 9 (2006): 513–18. https://doi.org/10.1136/jme.2005.013789.

Montessori, Maria. *The Absorbent Mind: A Classic in Education and Child Development for Educators and Parents.* New York: Holt Paperbacks Reprint Edition, 1995.

Montessori, Maria, and R. C. Orem. *A Montessori Handbook: "Dr. Montessori's Own Handbook."* New York: Putnam, 1966.

Morrison, J. "Human Nose Can Detect 1 Trillion Odours." *Nature,* March 2014. https://doi.org/10.1038/nature.2014.14904.

Mrug, Sylvie, et al. "Emotional and Physiological Desensitization to Real-Life and Movie Violence." *Journal of Youth and Adolescence* 44, no. 5 (2014): 1092–1108. https://doi.org/10.1007/s10964-014-0202-z.

Netburn, Deborah. "The Human Eye Can Detect a Single Photon, Study Finds." *Los Angeles Times,* July 19, 2016. https://www.latimes.com/science/sciencenow/la-sci-sn-human-eye-photon-20160719-snap-story.html.

"Nobel Laureate Orhan Pamuk of Turkey Is a Synesthete." *Sensorium* (blog). *Psychology Today.* https://www.psychologytoday.com/us/blog/sensorium/201306/nobel-laureate-orhan-pamuk-turkey-is-synesthete.

Nye, Bill. "Is the Human Species Still Evolving?" *Popular Science,* April 26, 2021. https://www.popsci.com/article/science/human-species-still-evolving/.

Ohsu, Takeaki, et al. "Involvement of the Calcium-Sensing Receptor in Human Taste Perception." *Journal of Biological Chemistry* 285, no. 2 (2010): 1016–22. https://doi.org/10.1074/jbc.m109.029165.

Olsson, Mats. "Behavioral and Neural Correlates to Multisensory Detection." *PNAS*. https://www.pnas.org/doi/10.1073/pnas.1617357114.

Padgett, Jason, and Maureen Seaberg. *Struck by Genius: How a Brain Injury Made Me a Mathematical Marvel.* Boston: Houghton Mifflin Harcourt, 2014.

Pelletier, Kenneth R. "The Chutzpah Factor in Altered States of Consciousness." *Journal of Humanistic Psychology* 17, no. 1 (1977): 63–73. https://doi.org/10.1177/002216787701700106.

"The Perfect Yellow." *Radiolab.* https://www.radiolab.org/episodes/211193-perfect-yellow.

Philipp, Sebastian T., et al. "Enhanced Tactile Acuity Through Mental States." *Nature News.* https://www.nature.com/articles/srep13549.

Piccolino, Marco, and Nicholas J. Wade. "Galileo's Eye: A New Vision of the Senses in the Work of Galileo Galilei." *Perception* 37, no. 9 (2008): 1312–40. https://doi.org/10.1068/p6011.

Porter, Jane. "The Neuroscience of Imagination." *Fast Company,* February 17, 2014. https://www.fastcompany.com/3026510/the-neuroscience-of-imagination.

Quigley, Elizabeth. "Parkinson's Smell Test Explained by Science." BBC News, March 20, 2019. https://www.bbc.com/news/uk-scotland-47627179.

Ramachandran, Vilayanur. "The Neurons That Shaped Civilization." TED. https://www.ted.com/talks/vilayanur_ramachandran_the_neurons_that_shaped_civilization.

"The Red of His E String." *Sensorium* (blog). *Psychology Today.*

https://www.psychologytoday.com/us/blog/sensorium
/201202/the-red-his-e-string.

Reddan, Marianne Cumella, Tor Dessart Wager, and Daniela
Schiller. "Attenuating Neural Threat Expression with Imagination." *Neuron* 100, no. 4 (2018). https://doi.org/10.1016/j
.neuron.2018.10.047.

Reichenbach, Tobias, and A. J. Hudspeth. "The Physics of Hearing: Fluid Mechanics and the Active Process of the Inner
Ear." *Reports on Progress in Physics* 77, no. 7 (July 2014).
https://iopscience.iop.org/article/10.1088/0034-4885/77/7
/076601.

"Research Suggests Enhanced Auditory and Cognitive Skills
for Meditators." *Hearing Review*, May 11, 2016. https:
//hearingreview.com/inside-hearing/research/research
-suggests-enhanced-auditory-cognitive-skills-meditators.

Sacks, Oliver. *Anthropologist on Mars*. New York: Vintage, 2014.

Sakai, Saburo, with Martin Caidin and Fred Saito. *Samurai! The
Personal Story of Japan's Greatest Living Fighter Pilot.* New
York: Ballantine Books, 1957.

Salinas, Joel. *Mirror Touch: Notes from a Doctor Who Can Feel
Your Pain*. New York: HarperOne, 2018.

Schiller, Rebecca. "Billie Eilish Talks Debut Record & Immersive
Album Experience Event: 'I Worked So Hard on This Little
Baby.'" *Billboard*, March 29, 2019. https://www.billboard
.com/music/pop/billie-eilish-talks-debut-record-immersive
-album-experience-8504722/.

Seaberg, Maureen. "Could Tetrachromats Hold the Key to Early
Coronavirus Detection?" *Medium*, February 21, 2020. https:
//medium.com/swlh/could-tetrachromats-hold-the-key-to
-early-coronavirus-detection-78179a79f57f.

Seaberg, Maureen. "Rise of the Machine Empaths." *Sensorium*
(blog). *Psychology Today*. https://www.psychologytoday

com /us /blog /sensorium /201508 /rise -the -machine -empaths.

Seaberg, Maureen. *Tasting the Universe: People Who See Colors in Words and Rainbows in Symphonies: A Spiritual and Scientific Exploration of Synesthesia.* With a foreword by W. C. Bushell. Pompton Plains, NJ: New Page Books, 2011.

Segal, David. "This Man Is Not a Cyborg. Yet." *New York Times,* June 1, 2013. https://www.nytimes.com/2013/06/02 /business/dmitry-itskov-and-the-avatar-quest.html.

"Sensory Experiences." Sensory Trust. https://www.sensorytrust .org.uk/about/sensory.

"Shamanic Synesthesia of the Kalahari Bushmen." *Sensorium* (blog). *Psychology Today.* https://www.psychologytoday.com /us/blog/sensorium/201202/the-shamanic-synesthesia-the -kalahari-bushmen.

Silberman, Steve. "Neurodiversity Rewires Conventional Thinking About Brains." *Wired,* April 16, 2013. https://www.wired .com/2013/04/neurodiversity/.

Skoff, Gabriella. "Quantum Superposition Bridges the Classic World." Project Q, April 7, 2020. https://projectqsydney.com /quantum-superposition-bridges-the-classic-world/.

Smith, Barry C. "We Have Far More Than Five Senses." You-Tube video, 5:35. Posted by *Aeon* Video, November 1, 2016. https://www.youtube.com/watch?v=zWdfpwCghIw.

Speed, Laura J., and Asifa Majid. "Superior Olfactory Language and Cognition in Odor-Color Synaesthesia." *Journal of Experimental Psychology: Human Perception and Performance,* 2017. doi: 10.1037/xhp0000469.

Spinella, Marcello. *A Relationship Between Smell Identification and Empathy.* https://www.researchgate.net/profile /Marcello-Spinella/publication/11116095_A_relationship _between_smell_identification_and_empathy/links

/02e7e52a6344ba71ee000000/A-relationship-between-smell
-identification-and-empathy.pdf.

"Spycraft and Synesthesia." *Sensorium* (blog). *Psychology Today.* https://www.psychologytoday.com/us/blog/sensorium
/201810/spycraft-and-synesthesia.

Stewart, Jude. *Roy G. Biv: An Exceedingly Surprising Book About
Color.* New York: Bloomsbury, 2013.

Stromberg, Joseph. "Nine Surprising Facts About the Sense of
Touch." *Vox,* January 28, 2015. https://www.vox.com/2015
/1/28/7925737/touch-facts.

"'Synesthesia Is in the Mind, Not the Brain'—Geoffrey Rush," *Sensorium*(blog),*Psychology Today*.https://www.psychologytoday
.com/us/blog/sensorium/201409/synesthesia-is-in-the-mind
-not-the-brain-geoffrey-rush.

"Synesthetes: 'People of the Future." *Sensorium* (blog). *Psychology Today.* https://www.psychologytoday.com/us/blog
/sensorium/201203/synesthetes-people-the-future.

"Synesthetic Sommelier." *Sensorium* (blog). *Psychology Today.* https://www.psychologytoday.com/us/blog/sensorium
/201302/the-synesthetic-sommelier.

Templeton, A. R. "Has Human Evolution Stopped?" *Rambam
Maimonides Medical Journal* 1 (2010). https://pubmed.ncbi
.nlm.nih.gov/23908778/.

Than, Ker. "Superhuman Hearing Possible, Experiments Suggest." *National Geographic,* May 4, 2021. https://www
.nationalgeographic.com/science/article/110516-people
-hearing-aids-ears-science.

Theise, Neil D., et al. "Liver from Bone Marrow in Humans."
Hepatology 32, no. 1 (2000): 11–16. https://doi.org/10.1053
/jhep.2000.9124.

Thurman, Robert. *Tibetan Book of the Dead.* New York: Random
House, 1993.

"Toward a Better Definition of Synesthesia." *Sensorium* (blog). *Psychology Today.* https://www.psychologytoday.com/us/blog/sensorium/201411/toward-better-definition-synesthesia.

Turin, Luca. "Vibrational Olfaction in Flies and Humans." *Flavour* 3 (2014). https://doi.org/10.1186/2044-7248-3-s1-k2.

Tyson, Peter. "Are We Still Evolving?" *Nova,* December 14, 2009. https://www.pbs.org/wgbh/nova/article/are-we-still-evolving/.

Underwood, Emily. "Scientists Discover a Sixth Sense on the Tongue—for Water." *Science* (2017). https://doi.org/10.1126/science.aan6904.

"Vincent van Gogh Was Likely a Synesthete." *Sensorium* (blog). *Psychology Today.* https://www.psychologytoday.com/us/blog/sensorium/201308/vincent-van-gogh-was-likely-synesthete.

"Was Marilyn Monroe a Synesthete?" *Sensorium* (blog). *Psychology Today.* https://www.psychologytoday.com/us/blog/sensorium/201202/was-marilyn-monroe-synesthete.

"We Need Imagination Now More than Ever." *Harvard Business Review,* February 1, 2021. https://hbr.org/2020/04/we-need-imagination-now-more-than-ever.

"What Lack of Affection Can Do to You." *Sensorium* (blog). *Psychology Today.* https://www.psychologytoday.com/us/blog/affectionado/201308/what-lack-affection-can-do-you.

"#WhatColorIsTheDress." *Sensorium* (blog). *Psychology Today.* https://www.psychologytoday.com/us/blog/sensorium/201502/whatcoloristhedress-0.

"Why Isn't the Sky Blue?" *Radiolab.* https://www.radiolab.org/episodes/211213-sky-isnt-blue.

"Wicked Case of Synesthesia." *Sensorium* (blog). *Psychology Today.* https://www.psychologytoday.com/us/blog/sensorium/201202/wicked-case-synesthesia.

Wilson, Brian. *I Am Brian Wilson*. Boston: Da Capo, 2016.

Winter, Caroline. "The Mind's Eye: Synesthesia Has Business Benefits." Bloomberg, January 9, 2014. https://www.bloomberg.com/news/articles/2014-01-09/the-minds-eye-synesthesia-has-business-benefits.

Wright, Robin. "I Have Something in Common with Marilyn Monroe—and You Might, Too." *The New Yorker*, August 31, 2017. https://www.newyorker.com/culture/culture-desk/i-have-something-in-common-with-marilyn-monroeand-you-might-too.

Wysocki, Charles J. Monell Chemical Senses Center. September 28, 2020. https://monell.org/charles-j-wysocki/.

Yeshe, Thubten, Jonathan Landaw, and Philip Glass. *Introduction to Tantra: The Transformation of Desire*. Boston: Wisdom Publications, 2014.

Zusak, Markus. *The Book Thief*. New York: Random House, 2006.

Notes

1. Olfactory Grammar

1. John P. McGann, "Poor Human Olfaction Is a Nineteenth Century Myth," *Science* 356, no. 6338 (May 2017), https://www.ncbi.nlm.nih.gov/pmc/articles/PMC5512720/.

2. Joy Milne, interview by the author conducted on February 1, 2022.

3. Elizabeth Quigley, "Parkinson's Smell Test Explained by Science," BBC News, March 20, 2019, https://www.bbc.com/news/uk-scotland-47627179.

4. David Howes, interview by the author conducted on February 23, 2022.

5. Chris Gilbert, "Musky Smell and Parkinson's Disease," *Psychology Today* (blog), February 23, 2018, https://www.psychologytoday.com/us/blog/heal-the-mind-heal-the-body/201802/musky-smell-and-parkinsons-disease.

6. Laura J. Speed and Asifa Majid, "Superior Olfactory Language and Cognition in Odor-Color Synaesthesia," *Journal of Experimental Psychology: Human Perception and Performance*, 2017, doi: 10.1037/xhp0000469.

7. Carrie Barcomb, interview by the author conducted on March 30, 2020.

8. "Her Nose Knows," *Sensorium* (blog), *Psychology Today*, November 3, 2015, https://www.psychologytoday.com/us/blog/sensorium/201511/her-nose-knows.

9. "The Green Flash: The Colorful Scientific Journey of Dr. Charles Wysocki," Monell Center (blog), *Monell Sensations*, http://blog.monell.org/02/22/the-green-flash/.

10. Nancie George et al., "Ten Incredible Facts About Your Sense of Smell," *Everyday Health*, October 3, 2014, https://www.everydayhealth.com/news/incredible-facts-about-your-sense-smell/.

11. J. Morrison, "Human Nose Can Detect 1 Trillion Odours," *Nature* (March 2014), https://doi.org/10.1038/nature.2014.14904.

12. Debra Pollicino-Ludwig, interview conducted by the author on January 26, 2021.

13. Lesley Roy, interview conducted by the author on April 13, 2022.

14. Luca Turin, interview conducted by the author on October 3, 2021.

15. Désirée Maßberg and Hanns Hatt, "Human Olfactory Receptors: Novel Cellular Functions Outside of the Nose," *Physiological Reviews* 98, no. 3 (2018): 1739–63, https://doi.org/10.1152/physrev.00013.2017.

2. Threshold of Imagination

1. Marco Piccolino and Nicholas J. Wade, "Galileo's Eye: A New Vision of the Senses in the Work of Galileo Galilei," *Perception* 37, no. 9 (2008): 1312–40, https://doi.org/10.1068/p6011.

2. Piccolino and Wade, "Galileo's Eye."

3. W. C. Bushell, interview conducted by the author on April 12, 2022.

4. Alipasha Vaziri, interview conducted by the author on October 30, 2020.

5. Deborah Netburn, "The Human Eye Can Detect a Single Photon, Study Finds," *Los Angeles Times*, July 19, 2016, https://www.latimes.com/science/sciencenow/la-sci-sn-human-eye-photon-20160719-snap-story.html.

6. Rebecca Holmes, "An Eye on Experiments That Make Quantum Mechanics Visible: *Aeon* Essays," *Aeon*, April 24, 2019, https://aeon.co/essays/an-eye-on-experiments-that-make-quantum-mechanics-visible.

7. Rebecca Holmes, interview conducted by the author on January 16, 2020.

8. Saburo Sakai with Martin Caidin and Fred Saito, *Samurai! The Personal Story of Japan's Greatest Living Fighter Pilot* (New York: Ballantine Books, 1957), 19.

9. Bill Bryson, *A Short History of Nearly Everything* (New York: Random House, 2016), 29–32.

10. Jason Padgett, interview conducted by the author on January 23, 2020.

11. Darold Treffert, interview conducted by the author on November 2, 2012.

12. Kristopher Jake Patten, interview conducted by the author on March 1, 2020.

13. Gabriella Skoff, "Quantum Superposition Bridges the Classic

World," Project Q, April 7, 2020, https://projectqsydney.com
/quantum-superposition-bridges-the-classic-world/.

3. Peregrinations Through Sensations

1. Daphne Maurer et al., "Synesthesia in Infants and Very Young Children," Oxford Handbooks Online, 2013, https://doi.org/10.1093/oxfordhb/9780199603329.013.0003.

2. "Nobel Laureate Orhan Pamuk of Turkey Is a Synesthete," *Sensorium* (blog), *Psychology Today,* June 5, 2013, https://www.psychologytoday.com/us/blog/sensorium/201306/nobel-laureate-orhan-pamuk-turkey-is-synesthete.

3. "A Wicked Case of Synesthesia," *Sensorium* (blog), *Psychology Today,* February 10, 2012, https://www.psychologytoday.com/us/blog/sensorium/201202/wicked-case-synesthesia.

4. Anna Harris, "Sensory School," *A Sensory Education* (London: Routledge, 2020), 65–82, https://doi.org/10.4324/9781003084341-5.

5. Gerald Lee Gutek, *The Montessori Method: The Origins of an Educational Innovation* (Lanham, MD: Rowman & Littlefield, 2004), 184.

6. Maria Montessori and R. C. Orem, *A Montessori Handbook: "Dr. Montessori's Own Handbook"* (New York: Putnam, 1966), Kindle edition, 25.

7. Helen Keller, "Three Days to See," *Atlantic Monthly,* January 1933, American Foundation for the Blind, https://www.afb.org/about-afb/history/helen-keller/books-essays-speeches/senses/three-days-see-published-atlantic.

8. "Steiner Online Library Spiritual Science for Human Evolution," *Man as a Being of Sense and Perception* (1958), https://steinerlibrary.org/Lectures/206/APC1958/index.html.

9. Jean A. Ayres, *Sensory Integration and the Child* (Torrance, CA: Western Psychological Services, 1979), 38.

10. Maria Montessori, *The Absorbent Mind: A Classic in Education and Child Development for Educators and Parents* (New York: Holt Paperbacks Reprint Edition, 1995), 183.

11. Immanuel Kant, *Critique of Pure Reason* (1787; Aegitas, 2016), 21, Kindle.

4. Taste Making

1. Carl Darling Buck, *A Dictionary of Selected Synonyms in the Principal Indo-European Languages* (Chicago: University of Chicago Press, 1949), 264.

2. Jaime Smith, interview conducted by the author on February 22, 2022.

3. "The Synesthetic Sommelier," *Sensorium* (blog), *Psychology Today*, February 7, 2013, https://www.psychologytoday.com/us/blog/sensorium/201302/the-synesthetic-sommelier.

4. Marcello Spinella, *A Relationship Between Smell Identification and Empathy*, June 2002, https://www.researchgate.net/profile/Marcello-Spinella/publication/11116095_ A_relationship_ between_smell_identification_and_empathy/links/02e7e52a-6344ba71ee000000/.

5. Marzi Pecen, interview conducted by the author on February 16, 2022.

6. Julie L. Hall, "When Narcissists and Enablers Say You're Too Sensitive," *Psychology Today*, February 21, 2021, https://www.psychologytoday.com/us/blog/the-narcissist-in-your-life/202102/when-narcissists-and-enablers-say-youre-too-sensitive.

7. Marie-Anne Suizzo, "Pleasure Is Good: How French Children Acquire a Taste for Life," https://theconversation.com/pleasure-is-good-how-french-children-acquire-a-taste-for-life-51949, January 4, 2016.

8. "Synesthetes: 'People of the Future,'" *Sensorium* (blog), *Psychology Today*, March 3, 2012, https://www.psychologytoday

.com/us/blog/sensorium/201203/synesthetes-people-the
-future.

9. Kenneth R. Pelletier, "The Chutzpah Factor in Altered States
of Consciousness," *Journal of Humanistic Psychology* 17,
no. 1 (1977): 63–73, https://doi.org/10.1177/002216787
701700106.

10. Takeaki Ohsu et al., "Involvement of the Calcium-Sensing
Receptor in Human Taste Perception," *Journal of Biological Chemistry* 285, no. 2 (2010): 1016–22, https://doi.org/10
.1074/jbc.m109.029165.

11. Buck, *A Dictionary of Selected Synonyms in the Principal
Indo-European Languages*, 264.

12. Emily Underwood, "Scientists Discover a Sixth Sense on the
Tongue—for Water," *Science,* 2017, https://doi.org/10.1126
/science.aan6904.

13. "Science & Nature—Human Body and Mind—Science of
Supertasters," BBC, September 17, 2014, https://www.bbc.co
.uk/science/humanbody/body/articles/senses/supertaster
.shtml.

14. Steve Silberman, "Neurodiversity Rewires Conventional
Thinking About Brains," *Wired,* April 16, 2013, https://www
.wired.com/2013/04/neurodiversity/.

15. Amanda Gardner, "Love Salt? You Might Be a 'Supertaster,'"
CNN, June 16, 2010, https://www.cnn.com/2010/HEALTH
/06/16/salt.taste/index.html.

5. The Perception Quotient

1. Ellen B. Braaten and Dennis Norman, "Intelligence (IQ) Testing," American Academy of Pediatrics, November 1, 2006, https:
//publications.aap.org/pediatricsinreview/article-abstract
/27/11/403/34094/Intelligence-IQ-Testing?autologincheck
=redirected%3FnfToken.

2. Daniel Goleman, *Emotional Intelligence* (New York: Bantam Books, 2020), 37–38.

3. Asifa Majid and Nicole Kruspe, "Hunter-Gatherer Olfaction Is Special," *Current Biology* 28, no. 3 (2018), https://doi.org/10.1016/j.cub.2017.12.014.

4. Peter Tyson, "Are We Still Evolving?," PBS, December 14, 2009, https://www.pbs.org/wgbh/nova/article/are-we-still-evolving/.

5. John Hawks, "Where Is Human Evolution Heading?," *US News & World Report*, July 24, 2008, https://www.usnews.com/science/articles/2008/07/24/where-is-human-evolution-heading.

6. Bill Nye, "Is the Human Species Still Evolving?," *Popular Science*, October 31, 2014, https://www.popsci.com/article/science/human-species-still-evolving/.

7. "The Brain Detects Disease in Others Even Before It Breaks Out," *ScienceDaily*, May 24, 2017, https://www.sciencedaily.com/releases/2017/05/170524084650.htm.

8. Diane Ackerman, *A Natural History of the Senses* (New York: Knopf Doubleday, 2011), Kindle edition, 307–8.

9. David Abram, *The Spell of the Sensuous: Perception and Language in a More-Than-Human World* (New York: Vintage Books, 2017), Kindle edition, 73.

10. Environmental Protection Agency, September 7, 2021, https://www.epa.gov/report-environment/indoor-air-quality.

11. Richard Corsi and Laura Klopfenstein, University of Texas at Austin, July 30, 2018, http://sites.utexas.edu/corsi/?cat=1.

12. Sylvia Mrug et al., "Emotional and Physiological Desensitization to Real-Life and Movie Violence," *Journal of Youth and Adolescence* 44, no. 5 (2014): 1092–1108, https://doi.org/10.1007/s10964-014-0202-z.

13. Carrie C. Firman, interview conducted by the author on March 16, 2022.

14. David Howes, interview conducted by the author on February 23, 2022.

15. Muhamed Sacirbey, interview by the author conducted on April 11, 2022.

6. Gandhi Neurons

1. Patricia Lynne Duffy, interview conducted by the author on March 7, 2022.

2. Nadia Bolognini, Elena Olgiati, Annalisa Xaiz, Lucio Posteraro, Francesco Ferraro, and Angelo Maravita, "Touch to See: Neuropsychological Evidence of a Sensory Mirror System for Touch," *Cerebral Cortex* 22, no. 9 (2011): 2055–64, https://doi.org/10.1093/cercor/bhr283.

3. Vilayanur Ramachandran, "The Neurons That Shaped Civilization," TED, January 4, 2010, https://www.ted.com/talks/vilayanur_ramachandran_the_neurons_that_shaped_civilization.

4. Shenbing Kuang, "Toward a Unified Social Motor Cognition Theory of Understanding Mirror-Touch Synaesthesia," *Frontiers*, May 31, 2016, https://www.frontiersin.org/articles/10.3389/fn.2016.00246/full.

5. Tracy Brower, "Empathy Is the Most Important Leadership Skill According to Research," *Forbes*, January 12, 2022, https://www.forbes.com/sites/tracybrower/2021/09/19/empathy-is-the-most-important-leadership-skill-according-to-research/?sh=5adad3ed3dc5.

6. Carolyn Hart, interview conducted by the author on April 7, 2022.

7. "Bearing Witness to Mirror Touch Synesthesia," *Psychology*

Today, May 10, 2018, https://www.psychologytoday.com/us
/blog/sensorium/201805/bearing-witness-mirror-touch
-synesthesia.

7. Synesthesia Is a Place

1. "Toward a Better Definition of Synesthesia," *Psychology Today,* November 3, 2014, https://www.psychologytoday.com/us/blog/sensorium/201411/toward-better-definition-synesthesia.

2. Robin Wright, "I Have Something in Common with Marilyn Monroe—and You Might, Too," *The New Yorker,* August 31, 2017, https://www.newyorker.com/culture/culture-desk/i-have-something-in-common-with-marilyn-monroeand-you-might-too.

3. Richard Cytowic, "Synesthesia: Phenomenology and Neuropsychology: A Review of Current Knowledge," *Psyche: An Interdisciplinary Journal of Research On Consciousness,* 1995.

4. Michael S. Ambinder et al., "Human Four-Dimensional Spatial Intuition in Virtual Reality," *Psychonomic Bulletin & Review* 16, no. 5 (2009): 818–23, https://doi.org/10.3758/pbr.16.5.818.

5. Jade McLeod, interview conducted by the author on May 27, 2022.

6. "The Shamanic Synesthesia of the Kalahari Bushmen," *Psychology Today,* February 15, 2012, https://www.psychologytoday.com/us/blog/sensorium/201202/the-shamanic-synesthesia-the-kalahari-bushmen.

7. "'Synesthesia Is in the Mind, Not the Brain'—Geoffrey Rush," *Psychology Today,* September 1, 2014, https://www.psychologytoday.com/us/blog/sensorium/201409/synesthesia-is-in-the-mind-not-the-brain-geoffrey-rush.

8. "The Red of His E String," *Psychology Today*, February 21, 2012, https://www.psychologytoday.com/us/blog/sensorium /201202/the-red-his-e-string.

9. "He's Got a Way About Him," *Psychology Today*, May 1, 2012, https://www.psychologytoday.com/us/blog/sensorium /201205/hes-got-way-about-him.

10. "Watch the Tonight Show Starring Jimmy Fallon Highlight: Billie Eilish Talks Happier than Ever, Directing Music Videos and Her Synesthesia," NBC, August 9, 2021, https: //www.nbc.com/the-tonight-show/video/billie-eilish -talks-happier-than-ever-directing-music-videos-and-her -synesthesia/246981503.

11. Rebecca Schiller, "Billie Eilish Talks Debut Record & Immersive Album Experience Event: 'I Worked so Hard on This Little Baby,'" *Billboard*, March 29, 2019, https://www .billboard.com/music/pop/billie-eilish-talks-debut-record -immersive-album-experience-8504722/.

12. "Was Marilyn Monroe a Synesthete?," *Psychology Today*, February 2, 2012, https://www.psychologytoday.com/us/blog /sensorium/201202/was-marilyn-monroe-synesthete.

13. "Vincent van Gogh Was Likely a Synesthete," *Psychology Today*, August 26, 2013, https://www.psychologytoday.com /us/blog/sensorium/201308/vincent-van-gogh-was-likely -synesthete.

14. Lynn Goode, interview conducted by the author on April 8, 2022.

15. Caroline Winter, Bloomberg, January 9, 2014, https://www .bloomberg.com/news/articles/2014-01-09/the-minds-eye -synesthesia-has-business-benefits.

16. "Hilma af Klint: Synesthete," *Psychology Today*, March 8, 2019, https://www.psychologytoday.com/us/blog/sensorium /201903/hilma-af-klint-synesthete.

17. Jim Channon, interview conducted by the author on July 7, 2015.

18. Doug Coupland, quoted in Maureen Ann Seaberg, *Tasting the Universe: People Who See Colors in Words and Rainbows in Symphonies: A Spiritual and Scientific Exploration of Synesthesia* (Pompton Plains, NJ: New Page Books, 2011), 251–52.

19. Roger Penrose, interview conducted by the author on September 2, 2009.

8. Touched by Sound

1. Y. Bitterman, R. Mukamel, R. Malach, I. Fried, and I. Nelken, "Ultra-Fine Frequency Tuning Revealed in Single Neurons of Human Auditory Cortex," *Nature News,* January 2008, https://www.nature.com/articles/nature06476.

2. W. C. Bushell, "Yogic Perception: An Integrative Scientific Model of Vajrayāna Meditational Practices," *Proceedings of the Second International Conference on Vajrayāna Buddhism* (Thimphu, Bhutan: Centre of Bhutan Studies, 2018), 130–55.

3. Tobias Reichenbach and A. J. Hudspeth, "The Physics of Hearing: Fluid Mechanics and the Active Process of the Inner Ear," *Reports on Progress in Physics* 77, no. 7 (July 2014), https://iopscience.iop.org/article/10.1088/0034-4885/77/7/076601.

4. Meredith LeMasurier and Peter G. Gillespie, "Hair-Cell Mechanotransduction and Cochlear Amplification," *Neuron* 48 (November 3, 2005): 403–15, https://doi.org/10.1016/j.neuron.2005.10.017.

5. Ker Than, "Superhuman Hearing Possible, Experiments Suggest," *National Geographic,* May 4, 2021, https://www.nationalgeographic.com/science/article/110516-people-hearing-aids-ears-science.

6. "The Extraordinary Sensorium of Brian Wilson," *Psychology*

Today, June 12, 2017, https://www.psychologytoday.com/us/blog/sensorium/201706/the-extraordinary-sensorium-brian-wilson.

7. Arla Good, Maureen Reed, and Frank Russo, "Compensatory Plasticity in the Deaf Brain: Effects on Perception of Music," *Brain Sciences* 4 (December 4, 2014): 560–74, https://doi.org/10.3390/brainsci4040560.

8. Martin D. Brazeau and Per E. Ahlberg, "Tetrapod-like Middle Ear Architecture in a Devonian Fish," *Nature* 439, no. 7074 (January 19, 2006): 318–21, https://doi.org/10.1038/nature04196.

9. Steve Roach, interview conducted by the author on March 16, 2022.

10. Dan Hasson, Töres Theorell, Martin Benka Wallén, Constanze Leineweber, and Barbara Canlon, "Stress and Prevalence of Hearing Problems in the Swedish Working Population—BMC Public Health," BioMed Central, February 23, 2011, https://bmcpublichealth.biomedcentral.com/articles/10.1186/1471-2458-11-130.

11. Psychiatric Neuroimaging Research, "Meditation Experience Is Associated with Increased Cortical Thickness," *Neuroreport,* November 28, 2005, https://journals.lww.com/neuroreport/Abstract/2005/11280/Meditation_experience_is_associated_with_increased.5.aspx.

12. "Is There Such a Thing as a Super Hearer?," Australian Broadcasting Corporation, July 25, 2012, https://www.abc.net.au/science/articles/2012/07/25/3553426.htm.

13. Kristopher Jake Patten, interview conducted by the author on March 1, 2022.

14. Randy Eady, interview conducted by the author on March 29, 2022.

15. Barry C. Smith, "We Have Far More Than Five Senses," You-

Tube video, 5:35, posted by *Aeon* Video, November 1, 2016, https://www.youtube.com/watch?v=zWdfpwCghIw.

16. "Our Sensory Approach." Sensory Trust, https://www.sensory trust.org.uk/about/sensory.

9. Infinitesimal Touch

1. David Howes, interview conducted by the author on February 23, 2022.

2. Alice Hines and Arden Wray, "Magnet Implants? Welcome to the World of Medical Punk," *The New York Times*, May 12, 2018, https://www.nytimes.com/2018/05/12/us/grindfest -magnet-implants-biohacking.html.

3. "What Lack of Affection Can Do to You," *Psychology Today*, August 31, 2013, https://www.psychologytoday.com/us/blog /affectionado/201308/what-lack-affection-can-do-you.

4. Margaret Atwood, *The Blind Assassin* (London: Virago, 2019), 308.

5. Dacher Keltner, *Dacher Keltner On Touch, Greater Good*, September 28, 2010, https://greatergood.berkeley.edu/video /item/dacher_keltner_on_touch.

6. Joseph Stromberg, "9 Surprising Facts About the Sense of Touch," *Vox*, January 28, 2015, https://www.vox.com/2015/1 /28/7925737/touch-facts.

7. Darren Lipomi, "(Invited) Organic Haptics: Soft Materials for Artificial Touch," *ECS Meeting Abstracts* MA2018-01, no. 26 (2018): 1540, https://doi.org/10.1149/ma2018-01/26/154.

8. Darren Lipomi, interview conducted by the author on January 25, 2022.

9. Lauren L. Orefice, "Outside-in: Rethinking the Etiology of Autism Spectrum Disorders," *Science* 366, no. 6461 (October 4, 2019): 45–46, https://doi.org/10.1126/science.aaz3880.

10. David Moore, interview conducted by the author on March 25, 2022.

11. "Sensation," Jack Westin, October 25, 2021, https://jackwestin .com/resources/mcat-content/sensory-processing/sensation.

12. Steven Pinker, interview conducted by the author on April 23, 2019.

13. Carol S. Dweck, *Mindset* (London: Robinson, 2017), 142.

10. The Highest Sense

1. Neil Theise, interview conducted by the author on March 25, 2002.

2. Walter Isaacson, "The Light-Beam Rider," *The New York Times*, October 30, 2015, https://www.nytimes.com /2015/11/01/opinion/sunday/the-light-beam-rider .html.

3. "Daniel's Eyes of China Blue," *Psychology Today*, May 11, 2012, https://www.psychologytoday.com/us/blog/sensorium /201205/daniels-eyes-china-blue.

4. "Dr. Neil Theise's Magnificent Time Wheels," *Psychology Today*, February 14, 2012, https://www.psychologytoday.com /us/blog/sensorium/201202/dr-neil-theises-magnificent -time-wheels.

5. Jane Porter, "The Neuroscience of Imagination," *Fast Company*, February 17, 2014, https://www.fastcompany.com/3026510 /the-neuroscience-of-imagination.

6. Marianne Cumella Reddan, Tor Dessart Wager, and Daniela Schiller, "Attenuating Neural Threat Expression with Imagination," *Neuron* 100, no. 4 (2018), https://doi.org/10.1016/j .neuron.2018.10.047.

7. Martin Reeves and Jack Fuller, "We Need Imagination Now More Than Ever," *Harvard Business Review*, February 1, 2021,

https://hbr.org/2020/04/we-need-imagination-now-more
-than-ever.

8. Paul Giamatti and Stephen T. Asma, "Imagination Is the Sixth Sense. Be Careful How You Use It: *Aeon* Essays," *Aeon*, March 2, 2022, https://aeon.co/essays/imagination-is-the-sixth-sense-be-careful-how-you-use-it.

9. Lesley Roy, interview conducted by the author on April 13, 2022.

11. The Eyes of Eve

1. "The Perfect Yellow," *Radiolab*, May 21, 2012, https://www.radiolab.org/episodes/211193-perfect-yellow.

2. Jay Neitz, interview conducted by the author on October 9, 2014.

3. A. R. Templeton, "Has Human Evolution Stopped?," *Rambam Maimonides Medical Journal*, National Library of Medicine, July 1, 2010, https://pubmed.ncbi.nlm.nih.gov/23908778/.

4. Kristopher Jake Patten, interview conducted by the author on March 1, 2022.

5. "#WhatColorIsTheDress," *Psychology Today*, February 27, 2015, https://www.psychologytoday.com/us/blog/sensorium/201502/whatcoloristhedress-0.

6. Anders Ericsson, *The Cambridge Handbook of Expertise and Expert Performance* (Cambridge, England: Cambridge University Press, 2006).

7. Joe Tsien, interview conducted by the author on December 1, 2016.

8. Markus Zusak, *The Book Thief* (New York: Random House, 2006), Kindle edition, 3–4.

9. G. Jordan, S. S. Deeb, J. M. Bosten, and J. D. Mollon, "The Dimensionality of Color Vision in Carriers of Anomalous

Trichromacy," *Journal of Vision* 10, no. 8 (2010): 12, https://doi.org/10.1167/10.8.12.

10. Veronique Greenwood, "The Humans with Super Human Vision," *Discover*, May 24, 2020, https://www.discovermagazine.com/mind/the-humans-with-super-human-vision.

11. Nikki A., "YInMn Blue: Rare Pigment That 'Absorbs Radiation' Is Selling at $179," *Tech Times*, January 25, 2021, https://www.techtimes.com/articles/256295/20210125/yinmn-blue-rare-pigment-absorbs-radiation-selling-179.htm.

12. Macrina Cooper-White, "Mystery of Dinosaur Feathers Finally Has a Solution," *HuffPost*, November 21, 2014, https://www.huffpost.com/entry/evolution-feathers-dinosaurs_n_6100020.

13. Klara Katarina Nordén, Chad M. Eliason, and Mary Caswell Stoddard, "Evolution of Brilliant Iridescent Feather Nanostructures," *eLife* 10 (2021), https://doi.org/10.7554/elife.71179.

14. Marie-Claire Koschowitz et al., "Beyond the Rainbow-Science," *Science*, https://www.science.org/doi/10.1126/science.1258957.

15. "Why Isn't the Sky Blue?," *Radiolab*, May 21, 2012, https://www.radiolab.org/episodes/211213-sky-isnt-blue.

16. M. Chirimuuta, *Outside Color: Perceptual Science and the Puzzle of Color in Philosophy* (Cambridge, MA: MIT Press, 2017), 131–58.

17. David Chalmers, interview conducted by the author on October 8, 2017.

12. The Senses, Meditation, and Consciousness

1. Sebastian Philipp, Tobias Kalisch, Thomas Wachtler, and Hubert R. Dinse, "Enhanced Tactile Acuity through Mental States," *Nature News*, August 27, 2015, https://www.nature.com/articles/srep13549.

2. "Research Suggests Enhanced Auditory and Cognitive Skills for Meditators," *Hearing Review,* May 11, 2016, https://hearingreview.com/inside-hearing/research/research-suggests-enhanced-auditory-cognitive-skills-meditators.

3. Tanuj Dada and Meghal Gagrani, "Mindfulness Meditation Can Benefit Glaucoma Patients," *Journal of Current Glaucoma Practice,* 2019, https://www.ncbi.nlm.nih.gov/pmc/articles/PMC6710928/.

4. Catherine J. Norris, Daniel Creem, Reuben Hendler, and Hedy Kober, "Brief Mindfulness Meditation Improves Attention in Novices: Evidence from Erps and Moderation by Neuroticism," *Frontiers in Human Neuroscience,* August 6, 2018, https://www.frontiersin.org/articles/10.3389/fnhum.2018.00315/full.

5. W. C. Bushell, interview conducted by the author on April 12, 2022.

6. David Segal, "This Man Is Not a Cyborg. Yet," *The New York Times,* June 1, 2013, https://www.nytimes.com/2013/06/02/business/dmitry-itskov-and-the-avatar-quest.html.

7. Robert Thurman, interview conducted by the author on April 9, 2022.

8. Maureen Seaberg, *Tasting the Universe: People Who See Colors in Words and Rainbows in Symphonies: A Spiritual and Scientific Exploration of Synesthesia* (Pompton Plains, NJ: New Page Books, 2011), 224–26.

9. Deepak Chopra, *Metahuman* (London: Penguin Random House UK, 2020), 14–15.

10. "Synesthesia: A Deep Look at Consciousness, ESP, Super-Normal Abilities and the Nature of Reality," YouTube video, 40:31, posted by Chopra Well, July 7, 2021, https://www.youtube.com/watch?v=9EH2LT3cCKg&t=1885s.

11. David Chalmers, "Facing Up to the Problem of Conscious-

ness," *Journal of Consciousness Studies* 2, no. 3 (1995): 200–19, http://www.consc.net/papers/facing.pdf.

12. "The Way of Grace," Sri Sri Ravi Shankar, https://www .srisri.com/p51/.

13. Eknath Easwaran, trans., *The Dhammapada*, Easwaran's Classics of Indian Spirituality Book 3 (Tomales, CA: Nilgiri Press, 2007), 244.

13. Frontiers

1. Vernor Vinge, "Technological Singularity," Carnegie Mellon Field Robotics Center, March 30, 1993, https://frc.ri.cmu.edu /~hpm/book98/com.ch1/vinge.singularity.html.

2. "Rise of the Machine Empaths," *Psychology Today*, August 30, 2015, https://www.psychologytoday.com/us/blog /sensorium/201508/rise-the-machine-empaths.

3. Jade McLeod, interview conducted by the author on June 15, 2022.

4. Robin May, interview conducted by the author on June 12, 2022.

5. Michelle Peck-Harris, interview conducted by the author on May 12, 2022.

6. Ray Kurzweil, *The Singularity Is Near: When Humans Transcend Biology* (London: Duckworth, 2013), 313.

7. Fritjof Capra, *The Tao of Physics: An Exploration of the Parallels Between Modern Physics and Eastern Mysticism* (Boston, MA: Shambhala Publications, 2010), 142.

8. M. J. McNamee and S. D. Edwards, "Transhumanism, Medical Technology and Slippery Slopes," *Journal of Medical Ethics* 32, no. 9 (2006): 513–8, https://doi.org/10.1136/jme .2005.013789.

9. Francis Fukuyama, "Transhumanism," *Foreign Policy*, no. 144 (2004): 42, https://doi.org/10.2307/4152980.

10. Zoltan Istvan, interview conducted by the author on June 21, 2022.

11. Edwin May, interview conducted by the author on September 15, 2021.

12. Joseph McMoneagle, interview conducted by the author on April 7, 2022.

14. Vivify Your Senses

1. Thubten Yeshe, Jonathan Landaw, and Philip Glass, *Introduction to Tantra: The Transformation of Desire* (Boston: Wisdom Publications, 2014), Kindle edition, 30–38.

2. Eden Phillpotts, *A Shadow Passes: Being the Third and Last Part of The Book of Avis* (New York: Macmillan, 1934), 17.

3. Luca Turin, interview conducted by the author on October 3, 2021.

Index